On-Lo

On-Location Recording Techniques

Bruce Bartlett
with
Jenny Bartlett

Focal Press

Boston Oxford Auckland Johannesburg Melbourne New Delhi

Focal Press is an imprint of Butterworth–Heinemann.

A member of the Reed Elsevier group

Recognizing the importance of preserving what has been written, Butterworth–Heinemann prints its books on acid-free paper whenever possible.

Butterworth–Heinemann supports the efforts of American Forests and the Global ReLeaf program in its campaign for the betterment of trees, forests, and our environment.

Library of Congress Cataloging-in-Publication Data
Bartlett, Bruce.
On-location recording techniques / Bruce Bartlett with Jenny Bartlett.
p. cm.
Includes bibliographical references and index.
ISBN 0-240-80379-5 (alk. paper)
1. Magnetic recorders and recording—Handbooks, manuals, etc.
I. Bartlett, Jenny. II. Title.
TK7881.6.B366 1999
621.389'32—dc21 99-10574
CIP

British Library Cataloguing-in-Publication Data
A catalogue record for this book is available from the British Library.

The publisher offers special discounts on bulk orders of this book.
For information, please contact:
Manager of Special Sales
Butterworth–Heinemann
225 Wildwood Avenue
Woburn, MA 01801-2041
Tel: 781-904-2500
Fax: 781-904-2620

For information on all Focal Press publications available, contact our World Wide Web home page at: http://www.focalpress.com

10 9 8 7 6 5 4 3 2 1

Printed in the United States of America

This book is fondly dedicated to the memory of
Mom, Dad, and Tom Lininger.

TABLE OF CONTENTS

13 Surround-Sound Techniques, DVD, and Super Audio CD *179*

14 Stereo Recording Procedures *195*

PREFACE

Sweat drips off the guitarist's face as he plunges into a solo. The audience yells and whistles their approval. In the back of his mind, the guitarist thinks, "I hope they're getting this on tape."

Perhaps the most exciting type of recording is to tape musicians playing "live" in a club or concert hall. Many bands want to be recorded in concert because they feel that is when they play best. Your job is to capture that performance on tape and bring it back alive.

Without a doubt, remote recording is exhilarating. The musicians—excited by the audience—often put on a stellar performance. You have only one chance to get it on tape, and it must be done right. You work on the edge. But by the end of the night, when everything has gone as planned, it is a great feeling.

This book, *On-Location Recording Techniques,* will help you do it right. It is the first book to focus exclusively on the special techniques used for recording outside the studio. It covers the unique requirements for capturing sound in a room or hall where the music is performed.

Whether you want to record an orchestra in a concert hall, a jazz combo in an auditorium, or a rock band in a club, you'll find useful tips here. If the band you're mixing wants to go on the road and would like you to do a mobile recording, you'll be able to do it. Some other remote applications are award shows, sporting events, radio broadcasts, and televised concerts.

The new breed of compact mixers and multitrack recorders has made going on location easier than ever. This book was written to help you take advantage of these new tools.

On-Location Recording Techniques is intended for recording engineers, record producers, musicians, broadcasters, and film-sound engineers—anyone who wants to know more about remote recording.

To understand this book, it helps to have a solid background in recording technology: microphones, mic techniques, connections, mixers, signal processors, and recorders. We assume that you know the basics of analog and digital recording. One introduction to recording technology is my book *Practical Recording Techniques* (2nd edition), published by Focal Press.

On-Location Recording Techniques is divided into two main parts: (1) popular music recording and (2) classical music recording. Let's look at Part 1 first.

Part 1. Popular Music Recording

Starting off Part 1, Chapter 1 offers a brief overview of recording equipment for taping pop music on location.

There are many ways to record live pop music, from simple to complex. Chapter 2 looks at each one. We also learn how to work with the sound-reinforcement system while making a recording.

Chapter 3 tells how to plan a live recording session. It covers preproduction meetings and site surveys. Also described is the paperwork needed to map out what you're going to do at the session. Based on my experience as an on-location recording engineer, Chapter 3 also offers tips for easier setup. Here you find shortcuts to make your job easier.

In Chapter 4 we go over the procedures at the actual recording session: connecting to power, running cables, miking, console setup, and so on. Chapter 5 suggests ways to mix and edit a live gig tape.

Finally, Chapter 6 describes how to build a quality recording truck. We look at truck acoustics, using as examples two high-end, commercial remote trucks.

With popular music, it is common to use multiple close mics and multitrack recorders. But with classical music, stereo mic techniques are the norm. Part 2 of *On-Location Recording* covers those methods.

Part 2. Classical Music Recording and Stereo Microphone Techniques

The heart of classical music recording is stereo mic techniques, so this topic is covered in great detail.

True stereo microphone techniques use two or three microphones to capture the overall sound of a sonic event. The stereo recording made from these microphones usually is reproduced over two speakers. The goal is to produce a believable illusion of the original performance and its acoustic environment in a solid, or three-dimensional, way.

There are many ways to make true stereo recordings, and Part 2 covers them all. It offers a clear, practical explanation of stereo miking theory, along with specific techniques, procedures, and hardware.

Stereo miking has several applications:

- Classical music recording of ensembles and soloists.
- Pop music recording: stereo pickup of piano, drums, percussion, and backing vocals.

- Stereo TV talk shows, game shows, audience reaction, electronic news gathering, sports, parades.
- Stereo film: feature film dialogue and ambience, documentaries.
- Stereo radio: group discussions and radio plays.
- Stereo sampling and sound effects.

For example, an orchestra might be recorded with two microphones and played back over two speakers. You would hear sonic images of the instruments in various locations between the stereo pair of speakers. These image locations—left to right, front to back—correspond to the instrument locations during the recording session. In addition, the concert hall acoustics are reproduced with a pleasing spaciousness. The result can be a beautiful, realistic recreation of the original event.

Part 2, Chapter 7, starts with an explanation of microphone polar patterns, which is necessary to understand how stereo techniques work. This is followed in Chapter 8 by a simple overview of the most common stereo microphone techniques.

Next, Chapter 9 covers stereo imaging theory in detail: how we localize real sound sources, how we localize images between loudspeakers, and how microphone techniques create images in various locations. We explore how to configure stereo arrays to achieve specific stereo effects. Spaciousness and spatial equalization are covered as well. This is the most academic chapter, offering a few mathematical equations related to the geometry of stereo miking.

Specific microphone techniques—such as XY, MS (mid-side), Blumlein, ORTF, OSS, SASS—are explained next: their characteristics, stereo effects, benefits, and drawbacks. Chapter 10 is devoted to free-field methods; Chapter 11 to boundary methods. The latest techniques are explained in detail, such as current developments in binaural and transaural stereo (Chapter 12).

Chapter 13 covers some new techniques for surround-sound miking. In a surround-sound recording of classical music, we usually hear the orchestra up front and the concert hall ambience from all around. Special mic techniques have been developed for capturing this surround effect. The main delivery mediums for surround recordings are DVD and Super Audio CD, which are described.

Armed with this knowledge, you're ready to record a musical ensemble. The necessary step-by-step procedures are described in Chapter 14. A troubleshooting guide helps you pinpoint and solve stereo-related problems.

Other stereo applications are explored in Chapter 15: television, video, film, nature recording, sports, sound effects, and sampling.

Part 2 concludes with a current listing of stereo microphones and accessories (Chapter 16). Each chapter in Part 2 includes a list of references for further reading.

Part 2 was based on my book *Stereo Microphone Techniques,* the first textbook written on stereo mic techniques, even though stereo miking has a history going back to 1881. It was first demonstrated at the International Exhibition of Electricity in Paris. A performance of the Paris Opera was picked up with several spaced pairs of microphones and transmitted 3 km away to people listening binaurally with two earphones (Hertz, 1981).

In the 1930s, British researcher Alan Blumlein and Bell Labs engineers Arthur Keller and Harvey Fletcher independently developed stereo technology for disk reproduction (Blumlein, 1958; Keller, 1981). Blumlein's patent in particular is a classic, far ahead of its time, and should be required reading for audio engineers. Stereo recording and reproduction became standard practice when stereo LPs were introduced in the late 1950s.

Our demands on stereo listening have become more sophisticated over the years. At first we were excited to hear simple left-right (ping-pong) stereo. Next we added a center image. Then we wanted accurate localization at all points, sharp imaging, and depth. Currently, methods are available to reproduce sound images all around the listener with only two loudspeakers in front or with 5.1 surround systems.

The need for up-to-date information on stereo microphone techniques has never been greater, and I trust that this book answers that need.

I hope you enjoy the thrill of live recording as much as I do.

References

A. Blumlein. "British patent specification 394,325." *Journal of the Audio Engineering Society* 6, no. 2 (April 1958), p. 91.

B. Hertz. "100 Years with Stereo: The Beginning." *Journal of the Audio Engineering Society* 29, no. 5 (May 1981), pp. 368–372.

A. Keller. "Early Hi Fi and Stereo Recording at Bell Laboratories (1931–1932)." *Journal of the Audio Engineering Society* 29, no.4 (April 1981), pp. 274–280.

All the above references can be found in *Stereophonic Techniques*, an anthology published by the Audio Engineering Society, 60 E. 42nd St., New York, NY 10165.

ACKNOWLEDGMENTS

Thank you to all the microphone manufacturers who sent photographs for this book. Thanks also to Terry Skelton, Mike Billingsley, Dan Gibson, and Ed Kelly for their suggestions on stereo miking.

I greatly appreciate the contributions and advice of reviewers Jim Loomis, an instructor at Ithaca College; Bruce Outwin, an instructor at Emerson College; and Ron Estes of NBC in Burbank. Thanks also to my charming and helpful editors at Focal Press: Marie Lee and Terri Jadick.

Thank you to the publishers who allowed me to use some of my own material for this book. Some chapters are reprinted with permission from

MR&M Publishing Corp. and Sagamore Publishing Co. Inc., Recording Techniques series by Bruce Bartlett.

Howard W. Sams & Co., *Introduction to Professional Recording Techniques,* Chapters 7 and 17, copyright 1987 by Bruce Bartlett.

Radio World, "Stereo Microphone Techniques Part 1" (November 1989) and "Stereo Microphone Techniques Part 2" (February 1990), by Bruce Bartlett.

Parts of Chapter 11 were based on the article by B. Bartlett, "An Improved Stereo Microphone Array Using Boundary Technology: Theoretical Aspects," *Journal of the Audio Engineering Society* 38, nos. 7–8 (July–August 1990), pp. 543–552.

TRADEMARKS

The following terms used in this book are believed to be trademarks or registered trademarks: Dolby, Dolby Digital, Dolby Surround, Dolby Stereo, Dolby Pro Logic, Dolby Surround Multimedia, AC-3, Mic-Eze, AudioReality, Cam-lok, Tube Traps, SRS, Sound Retrieval System, Qsys, RSS, Roland Sound Space, No-Noise, VR^2, Virtual Reality Recording, Digital Theater Systems, DTS, RPG, Syn Aud Con, Sonic Studio, Spatializer, Circle Surround, DSM,THX, Sonex, ITE, PAR, Pressure Recording Process, Stereo Ambient Sampling System, PCC, Phase Coherent Cardioid, Pressure Zone Microphone, PZM, SASS, SRS TruSurround, Q Surround, Dolby Virtual Surround, Sony Virtual Enhanced Surround, Panasonic Virtual Sonic, VLS Cyclone 3D, Sony DSD (Direct Stream Digital), Sony Super Bit Mapping Direct (SBM Direct), Sony Super Audio CD, Sorbothane.

Part 1
Popular Music Recording

1

EQUIPMENT FOR REMOTE RECORDING OF POPULAR MUSIC

Recording a band playing live on location is a great way to capture its musical energy. This book offers tips on making excellent remote recordings, whether for popular or classical music.

To understand the book, you need a solid background in recording technology: microphones, mic techniques, connectors, mixers, signal processors, and recorders. It helps to know the basics of analog and digital recording. One introduction to recording technology is my book, *Practical Recording Techniques* (2nd edition), published by Focal Press.

This chapter offers a brief overview of recording equipment for taping pop music on location. It is a simplified introduction for beginners and readers who want a review. Further reading is recommended. At the end of this book is a Glossary, which defines tech terms.

The Recording Chain

Figure 1–1 shows the parts of a typical multitrack recording system. Let's look at the signal flow from start to finish:

1. Musical instruments produce sound.
2. Microphones change sound into electrical signals.
3. Mic choice and placement affect the tone quality and the amount of room acoustics and leakage picked up.

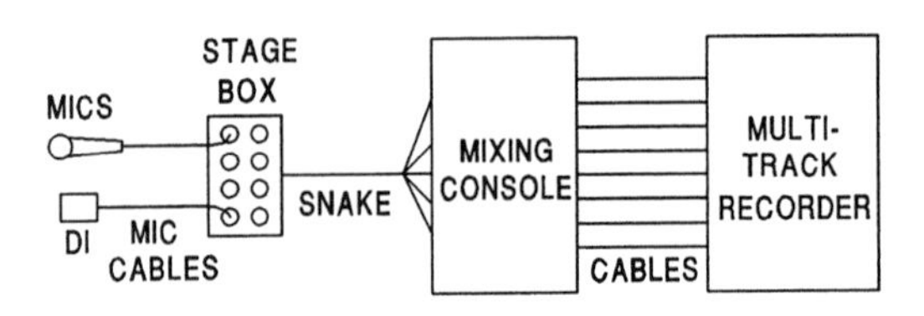

Figure 1–1
Signal flow in a typical multi-track recording system.

4. Mic cables carry the mic signals.
5. The mic cables plug into a stage box—a box with multiple mic connectors. Wired to the connectors is a multiconductor cable called a *snake.*
6. The snake connectors plug into a mixing console. If the musical event is reproduced over a sound system, the mixing console is the one used for sound reinforcement. It amplifies each mic signal up to a higher voltage called line level.
7. The amplified signal of each mic is available at an insert send jack on the mixer.
8. Cables plugged into the insert send jacks carry the signal to your multitrack tape recorder.
9. The multitrack unit records each mic's signal on a different track. You mix these tracks later, back in your studio.

The mixer signals might go first to a recording mixer (to set levels), and then to the multitrack recorder.

There are many other ways to record pop music on location, and this book covers them. For now, let's look at each part of the recording chain just described.

Microphones

A microphone changes sound into electricity. Based on how this is done, two main types of microphone are used for on-location recording: condenser and dynamic.

A condenser mic typically gives a clear, detailed, natural sound. It works great on cymbals, acoustic guitar, acoustic piano, and vocals. Condenser mics require a power supply to work, such as a battery, a separate phantom power supply, or phantom power built into a mixing console. In the console, phantom is turned on with a switch, either one switch for all the mics or individual switches for each mic.

A dynamic (moving coil) mic works with no power supply. It is rugged and reliable. Most dynamics do not sound as clear and natural as condensers. But dynamics with a "presence peak," a rise in the frequency response around 5 kHz, still are a popular choice for guitar amps, drums, and vocals.

Microphones also differ in the way they respond to sounds coming from different directions. An omnidirectional mic picks up sound equally well in all directions. It is seldom used for pop music recording on location because it picks up too much feedback from the PA system and too much leakage, sound from instruments other than the one at which the mic is aimed.

A unidirectional mic picks up sound best in front of the microphone. It partly rejects sounds to the sides and rear of the mic. For this reason, unidirectional mics are used to reduce feedback and leakage. Three types of unidirectional mic are cardioid, supercardioid, and hypercardioid. Each has a progressively more "focused" pickup pattern.

A stereo mic has two mic capsules in the same housing for convenient stereo recording.

Mics for pro recording are low impedance balanced. A low-impedance mic (under 600 ohms) lets you run long cables without hum pickup or high-frequency loss. Mics with a balanced output have an XLR-type (three-pin) connector.

A mic cable has two conductors surrounded by a shield. On one end of the cable is a female XLR-type connector (with three holes); on the other is a male XLR-type (with three pins). Pin 1 is shield, pin 2 is audio in-phase (hot), and pin 3 is audio opposite polarity (cold).

You plug the mic cables into a stage box. This is a metal chassis with several female XLR-type connectors (Figure 1–2). Each connector has a number. The box is wired to a single multiconductor cable, a snake. At the far end of the snake, the cable divides into several individual cables, each with a correspondingly numbered male XLR-type connector. These male XLRs plug into your mixer mic inputs.

A direct box (DI) replaces a microphone on electric instruments. The DI converts a signal from an electric guitar, electric bass, or synthesizer into a low-impedance balanced signal, which you plug into a mic input on the stage box. (An electric guitar amp normally is miked—not recorded direct—in order to capture the amp's distortion.)

Finally, foam pop filters or windscreens are placed on vocal mics to reduce explosive breath sounds.

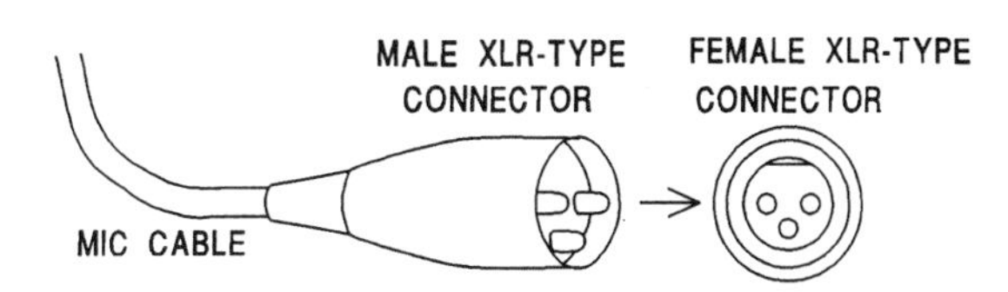

Figure 1–2
Male and female XLR-type connectors.

Mic Technique Basics

Microphone placement has a big effect on the sound of a recording. The farther a microphone is from its sound source, the more the microphone picks up feedback, room acoustics, background noise, and leakage (unwanted sound from other instruments). So when you record popular music during a concert, mike close, within a few inches, to reject these unwanted sounds. When you record classical music, mike farther away (4–20 feet) to pick up the hall reverberation, which is a desirable part of the sound.

Microphone placement also affects the tonal balance of a recorded instrument. When you change the mic position, you change the tone quality. For example, an acoustic guitar miked near the sound hole is bassy, near the bridge is mellow, and near the fingerboard is bright.

If you are recording a concert that is being reinforced through a sound system, you usually are forced to mike close to the loudest part of the instrument because of feedback and leakage. For example, in an acoustic guitar the loudest part is the sound hole. In a sax it's the bell, and so on. The tone quality is not natural in these spots, so you will have to use equalization to compensate. In the studio you are freer to experiment and find a miking spot that sounds good.

Multitrack Recorders

The multitrack tape machine records several tracks on magnetic tape. One track might be a lead vocal, another track might be a saxophone, and so on.

Analog multitrack recorders and some digital multitrack machines record on open reel tape. These recorders and tape are expensive. Less costly are modular digital mulitracks (MDMs), which record on videocassettes (Figure 1–3). For example, the Alesis ADAT XT20 records on a VHS cassette; the TASCAM DA-38 records on a Hi-8 videocassette.

Each MDM is a module containing eight tracks. By connecting multiple modules together, you can record 16, 24, or more tracks all in sync. The

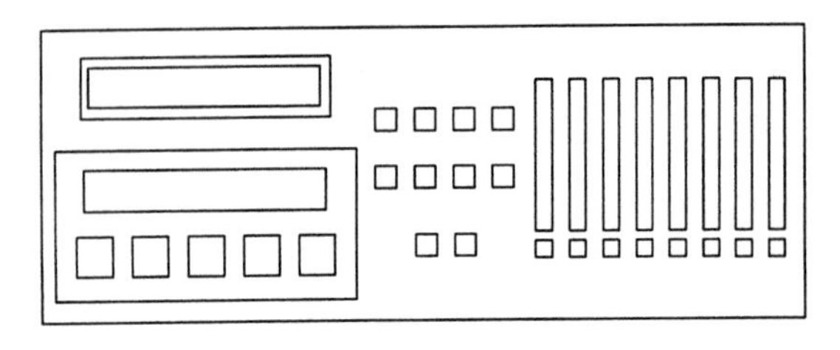

Figure 1–3
A modular digital multitrack (MDM) tape recorder.

TASCAM MDMs record up to 1 hour 48 minutes nonstop on a single cassette, making them ideal for recording long concerts.

Another option is a recorder-mixer or "portable studio." It combines a simple mixer and multitrack recorder in a single package. The recorder is analog (cassette) or digital (hard disk or MiniDisc). For on-location work, you need to be able to record on all tracks at once, so make sure the recorder-mixer can do this.

Yet another recording device is a multitrack hard disk recorder. Currently, this device can record only eight tracks for about 35 minutes. This may not be long enough for recording a concert. Also, at concerts you need to swap the recording medium during an intermission. This can be difficult or impossible with a hard disk recorder. For these reasons, multitrack tape recorders are preferred for on-location recording.

Often it's hard to control the recording level during a live concert. Sometimes the level may be quite low, resulting in audible hiss if you use an analog multitrack. So a digital multitrack tape recorder is usually best for on-location work.

Mixing Console

The mixing console (board or desk) is an audio control panel made of several identical modules, side by side. Each mic plugs into its own module. Here are the main controls on each input module (Figure 1–4):

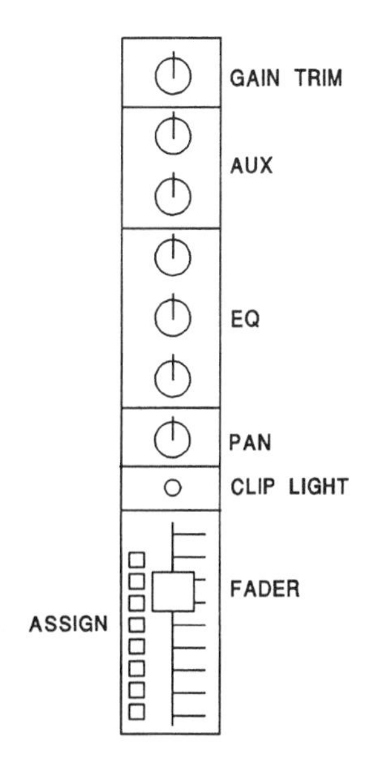

Figure 1–4
Typical input module in a mixing console.

- **Input trim or gain:** Sets the signal level from each mic to the optimum level. This prevents noise and distortion.
- **Fader:** A sliding volume control for each microphone.
- **EQ or equalization:** Tone control. Bass controls low frequencies (roughly 20–150 Hz); midbass controls 150–500 Hz; midrange controls middle frequencies (500 Hz–5 kHz); treble controls high frequencies (5–20 kHz).
- **Pan:** Sets the position of each instrument between your speakers—left, right, center, and so forth.
- **Aux:** Normally used to control the amount of effects (reverb, echo, chorus, etc.) on each tape track, but also can be used to set up a separate mix.
- **Assign:** Lets you route or send each mic signal to the desired output channel or bus.

Console Connections

This section describes how to connect a multitrack recorder to a mixing console. In making a multitrack remote recording, you generally want to record the signal of each mic on a different tape track. Those tracks will be mixed later in the studio.

If you try to plug a mic into a multitrack tape recorder, the recorder meters won't move because the signal level is too low. A recorder needs to be fed a line-level signal. This signal comes from your mixing console.

In the console are several mic preamplifiers, one per mic, which amplify the mic-level signal up to line level. For each mic channel, this line-level signal typically appears at two connectors on the back of the mixer: *direct out* and *insert send*. That's where to connect to the tape-track inputs.

Usually the *insert send* is the best connector to use. Here is why. Typically, the direct-out signal is postfader (Figure 1–5). This means the signal at the direct-out jack comes after the fader, so the signal is affected

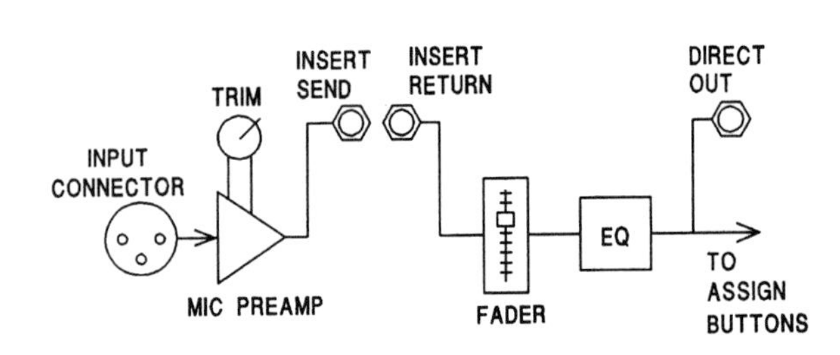

Figure 1–5
Simplified signal flow through part of a mixing console, showing insert and direct out.

by the fader (volume) settings. Any fader movements will show up on your tape, which is undesirable. It's better to connect tape tracks to insert sends. These usually are prefader, so moving the fader does not affect the recording level.

As the musicians are playing a loud song, adjust the trim or gain controls on the mixer to get about a –10 dB level on the recorder meters. On a digital recorder, 0 dB is the maximum level, so –10 dB allows for surprises. Once the trims are set, the PA operator can adjust the mix (move the faders) without affecting your recording levels.

In short, use the trim controls to set recording levels. Use the faders to set up a monitor mix or PA mix.

More on Insert Jacks

Some consoles have an insert-send jack and an insert-return jack. Each insert send connects to a tape-track input. Each insert return connects to a tape-track output. You must monitor the *input* signal on your multitrack recorder to hear a signal through the mixing console.

Some insert jacks use a single connector for send and return. The tip connection is the send, the ring is the return, and the sleeve is the common ground or shield.

Insert jacks also let you plug a compressor in series with an input module's signal for automatic volume control.

Assign Buttons

The *assign* buttons let you record several instruments on one track. Suppose you want to record all the drum mics on tracks 1 and 2. In the input module for each drum mic, press the Assign 1 and Assign 2 buttons. Then the drum mic signals will go to output busses 1 and 2, which you connect to the tape tracks 1 and 2 inputs.

Using the faders, set up a drum mix. Pan each mic as desired. Set the overall drum-mix level with the bus 1 and 2 faders (also called *group faders* or *submasters*).

Now that we've covered the basics, we can get into actual recording techniques—covered next.

2

RECORDING TECHNIQUES FROM SIMPLE TO COMPLEX

There are several ways to record pop music concerts:

- Record with two mics out front into a two-track recorder.
- Using the PA mixer, record off a spare main output.
- Get a four-track recorder-mixer. Place a stereo mic (or a pair of mics) at the FOH (front of house, the PA mixer location) position. Record the mic on tracks 1 and 2. Also plug into a spare main output on the PA mixer, and record it on tracks 3 and 4.
- Feed the PA mixer insert jacks to a multitrack tape recorder.
- Feed the PA mixer direct outs or insert jacks to a recording mixer, and from there to a two-track or a multitrack tape recorder.
- Use a mic splitter on stage to feed the PA snake and recording snake. Record to multitrack or two track.
- Do the multitrack recording in a van or mobile recording truck.

We start by explaining simple two-microphone techniques and work our way up to elaborate multitrack setups.

Two Mics out Front

Let's start with the simplest, cheapest technique: two mics and a two-track recorder. The sound will be distant and muddy compared to using a mic on each instrument and vocal. Not exactly CD quality. But you'll capture how the band sounds to an audience.

Recording this way is much simpler, faster, and cheaper than multi-mic, multitrack recording. Still, if time and budget permit, you'll get better sound with a more elaborate setup.

Equipment

Here is what you need for two-mic recording:

- Two mics of the same model number. Your first choice might be cardioid condenser mics. The cardioid pickup pattern cuts down on room reverb (reverberation) and noise. The condenser type generally sounds more natural than the dynamic type. Another option is a pair of boundary mics such as PZMs. Simply put them on the floor or on the ceiling in front of your group.
- A DAT recorder, two-track open-reel recorder, or cassette deck. DAT has the least distortion and hiss, and lets you record up to two hours nonstop. Use a unit with either mic inputs or a separate mic preamp (preamplifier).
- Blank recording tape. Buy the best you can afford. Bring enough tape to cover the duration of the recording.
- Two long mic cables.
- Two mic stands.
- Headphones. You could use speakers in a separate room, but headphones are more portable and they sound consistent in any environment. Closed-cup headphones partly block out the live sound of the band so you can better hear what's going on tape. Ideally, you'd set up in a different room than the band, so you can clearly hear what you're recording.

Mic Placement for a Band Using a PA System

Use a pair of mic stands or hang the mics out of the reach of the audience. Aim the two mics at the group about 12 feet away and space them about 5–15 feet apart (Figure 2–1). Place the mics far apart (that is, close to the PA speakers) to make the vocals louder in the recording. Do the opposite to make them quieter. The stereo imaging will be vague, but at least you can control the balance between instruments and vocals.

Mic Placement for a Band Not Using a PA System

To record a small folk group or acoustic jazz group without a PA, set up two mics of the same model number in the stereo arrangement of your

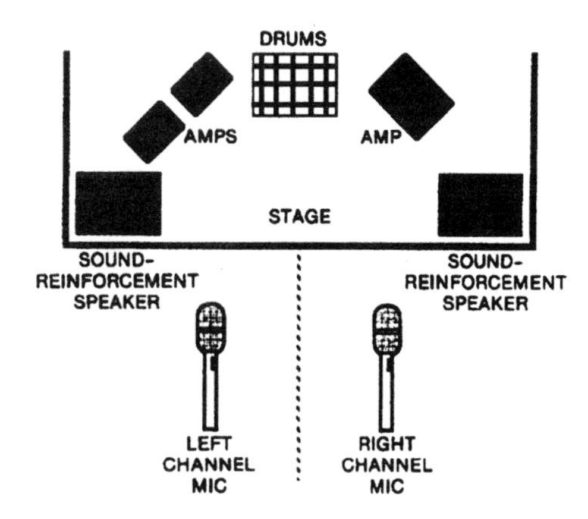

Figure 2–1
Recording a musical group with two equally spaced microphones.

choice. Place the mics about 3–6 feet from the group. The balance may not be the best, but the method is simple. Details are found in Chapter 14 under the heading "Stereo Miking for Pop Music." Chapters 10–13 describe specific stereo mic arrays.

Recording

After setting the recording level, leave it alone as much as possible. If you must change the level, do so slowly and try to follow the dynamics of the music. Switch tapes during pauses or intermissions.

If the playback sounds distorted—even though you did not exceed a normal recording level—the mics probably overloaded the mic preamps in the tape deck. A mic preamp is a circuit in the tape recorder that amplifies

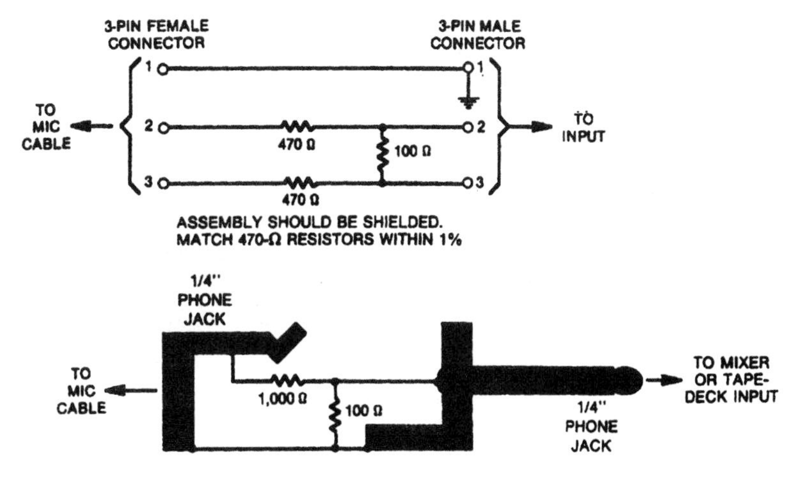

Figure 2–2 Balanced and unbalanced microphone pads.

a weak mic signal up to a usable level. With loud sound sources such as rock groups, a mic can put out a signal strong enough to cause distortion in the mic preamp.

Some recorders have a pad or input attenuator. This reduces the mic signal level before it reaches the preamp and so prevents distortion. You can build a pad (Figure 2–2) or buy some plug-in pads from your mic dealer. Some condenser mics have a switchable internal pad that reduces distortion in the mic itself. If you have to set your record-level knobs very low (less than $\frac{1}{3}$ up) to get a 0 recording level, that shows you probably need to use a pad.

Recording from the PA Mixer

You can get a fairly good recording by plugging into the main output of the band's PA mixer. (*PA* stands for public address, but here it means sound reinforcement.) Connect the main output(s) of the mixer to the line or aux input(s) of a two-track recorder. Use the mixer output that is ahead of any graphic equalizer used to correct the speakers' frequency response (Figure 2–3).

Mixers with balanced outputs can produce a signal that is too high in level for the recorder's line input, causing distortion. This probably will occur if your record-level controls have to be set very low. To reduce the output level of the mixer, turn it down so that its signal peaks around –12 VU on the mixer meters, and turn up the PA power amplifier to compensate. That practice, however, degrades the mixer's S/N (signal-to-noise) ratio.

A better solution is to make a 12 dB pad (Figure 2–4). The output level of a balanced-output mixer is 12 dB higher than the normal input level of a recorder with an unbalanced input.

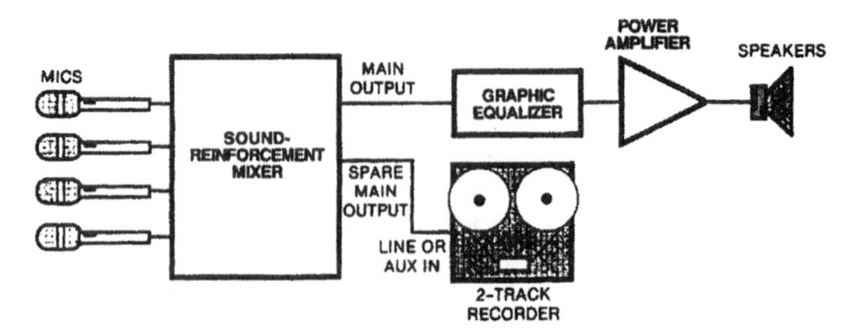

Figure 2–3 Recording from the sound-reinforcement mixer.

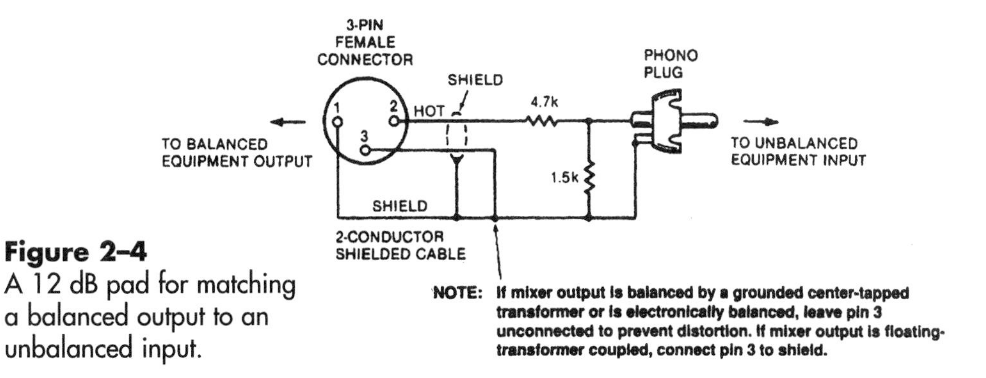

Figure 2–4
A 12 dB pad for matching a balanced output to an unbalanced input.

Drawbacks

The recorded mix off the PA might be poor. The operator of the band's mixer hears a combination of the band's live sound and the reinforced sound through the house system and tries to get a good mix of both these elements. That means the signal is mixed to augment the live sound, not to sound good by itself. A recording made from the band's mixer is likely to sound too strong in the vocals and too weak in the bass.

Taping with a Four Tracker

A four-track portable studio can do a good job of capturing a band's live sound. With this method, you place a stereo mic (or a pair of mics) at the FOH position. Record the mic on tracks 1 and 2. Also plug into a spare main output on the PA mixer, and record it on tracks 3 and 4 (Figure 2–5). After the concert, mix the four tracks together. You may be surprised at the quality you get with this simple method.

The FOH mics pick up the band as the audience hears it: lots of room acoustics, lots of bass, but rather muddy or distant. The PA mixer output sounds close and clear but typically is thin in the bass. When you mix the

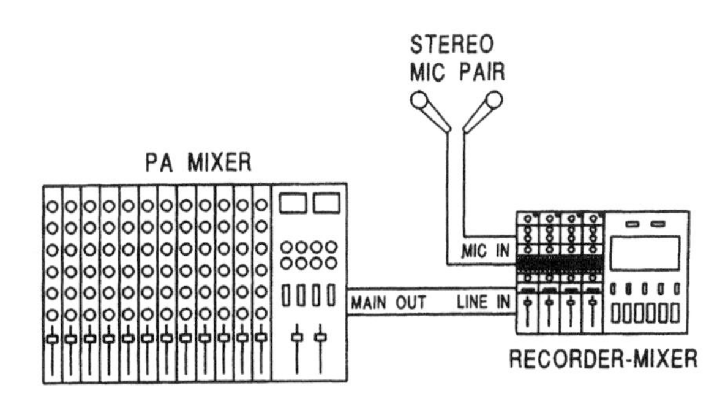

Figure 2–5
Recording two mics and a PA mix on a four-track recorder-mixer.

FOH mics with the PA mixer signal, the combination has both warmth and clarity.

Consider using a stereo mic at the FOH position. Good stereo mics cost over $500, but they provide great stereo in a portable, convenient package. You also can set up two mics of the same model number in a stereo arrangement. For example, angle two cardioid mics 90° apart and space them 1 foot apart horizontally. Or place two omni mics 2 feet apart.

Plug the mics into your four track's mic inputs 1 and 2. Adjust the trim to prevent distortion, and set the recording level. Find a spare main output on the PA mixer, and plug it into your four track's line inputs 3 and 4. You may need to use the 12 dB pad described earlier. Set recording levels and record the gig.

Back home, mix the four tracks to stereo. Tracks 1 and 2 provide ambience and bass; tracks 3 and 4 provide definition and clarity.

You might hear an echo because the FOH mics pick up the band with a delay (sound takes time to travel to the mics). A typical delay is 20–100 msec. To remove the echo, delay the PA mix by the same amount. Patch a delay unit into the access jacks for tracks 3 and 4. As you adjust the delay time up from 0, the echo will get shorter until the signals are aligned in time.

Recording from the PA Mixer Aux Output

On the PA mixer, find an unused aux send output. Plug in a Y-cord: One end fits the PA mixer connector; the other end has two connectors that mate with your two-track recorder's line input (Figure 2–6). If you have two spare aux busses, you could plug a cable into both of them and set up a stereo mix.

Put on some good closed-cup headphones and plug them into your two-track recorder to monitor the recording. Adjust the aux-send knob for each instrument and vocal to create a good recording mix. Record the gig.

The advantages of this method are

- It is simple. All you need is a recorder, a cable, and headphones.

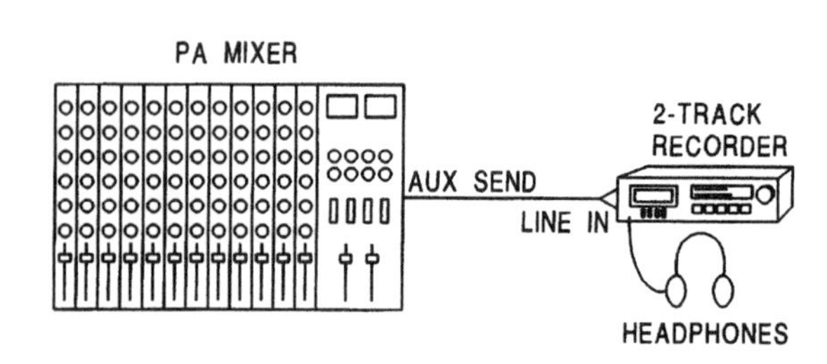

Figure 2–6
Recording from the PA mixer aux outputs.

- The recorded sound is close and clear.
- If the mix is done well, the sound quality can be very good.

The disadvantages are

- It is hard to hear what you're mixing. You may need to do several trial recordings. Set up a mix, record, play back, and evaluate. Redo the mix and try another recording. It also helps to use in-the-ear phones covered by industrial hearing protectors.
- As you adjust the aux knobs, you might get in the way of the PA operator.
- The recording will be dry (without effects or room ambience). However, you could plug two room mics into the PA mixer and add them to the recording mix. Do not assign these mics to the PA output channels.
- If the aux send is pre-EQ, there will be no EQ on the mics. If the aux send is post-EQ, there will be EQ on the mics, but it may not be appropriate for recording.

Recording an aux mix works best where the setup is permanent and you have time to experiment. Some examples are recording a church service or recording a regularly scheduled show in a fixed venue.

Feed the PA Mixer Insert Sends to a Modular Digital Multitrack

This is an easy way to record, and it offers very good sound quality with minimal equipment. Plug one or more MDMs into the insert-send jacks on the back of the PA mixer. Set levels with the PA mixer input trims. After the concert, mix the tape tracks back in your studio.

This method has some drawbacks. The sound quality depends on the PA mixer's preamps. Changes in the mixer input trims show up on your tape. You may have to ask the PA operator to adjust the input trims during the concert. If some insert jacks are tied up with signal processors, you must use those channels' direct-out jacks instead, which are usually post-fader (unless they can be switched to prefader).

Connections

First, find out what kind of insert jacks the PA mixer has and what kind of unbalanced input connectors your MDM has. Buy or make some

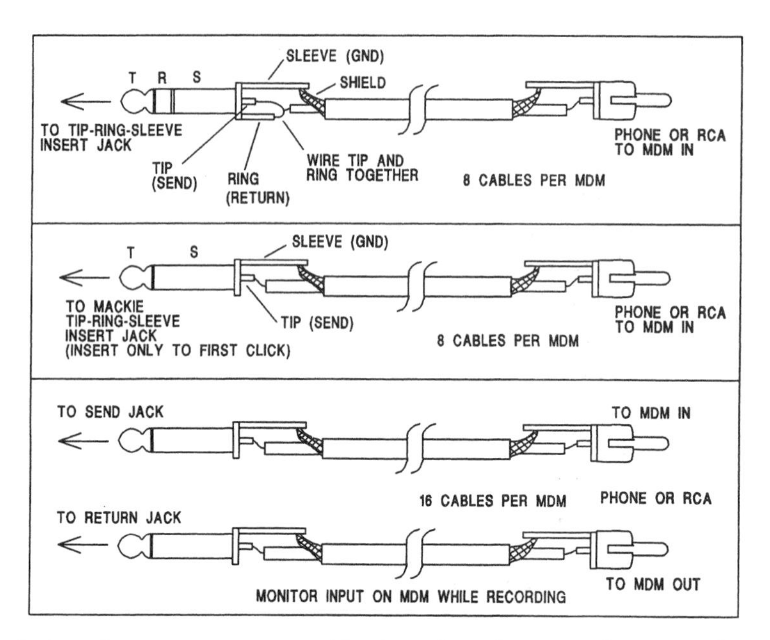

Figure 2–7 Three ways to wire cables based on the type of insert jack.

single-conductor shielded cables that mate with those connectors. Figure 2–7 shows three ways to wire cables based on the type of insert jack.

Suppose you want to record one instrument or vocal on each track. In each PA mixer input channel, locate the insert or access jack. Connect that jack to a tape-track input (Figure 2–8). Insert jacks usually are prefader, pre-EQ. So any fader or EQ changes done by the PA operator will not show up on your tape. (Chapter 1 explains this concept.) However, any changes the PA operator makes in the trim settings during the show will affect your recording as well.

If the PA mixer has separate send and receive insert jacks, connect the send to the tape-track input and connect the receive to the tape-track output. Set your MDM to monitor the input signal so that the PA mixer will receive a signal.

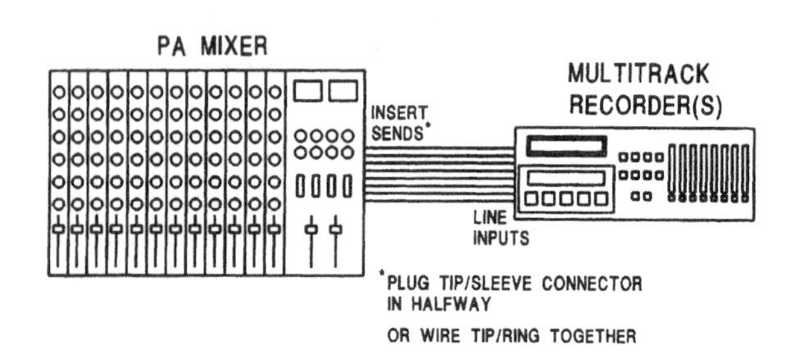

Figure 2–8
Feeding an MDM from the PA mixer insert jacks.

Some boards use a single stereo insert jack with TRS (tip/ring/sleeve) connections. Usually the tip is send and the ring is receive. In the stereo phone jack you plug into the insert jack, wire the tip and ring together and also to the cable hot conductor. That way, the insert-send goes directly to the insert-receive and also to the recorder.

Another option is to use a snake with tip/ring/sleeve stereo phone plugs at the console end. Carry some TRS-to-dual-mono phone adapters to handle consoles that have separate jacks for insert send and receive.

On some PA mixers with a tip/ring/sleeve insert jack, you can use a mono (tip/sleeve) phone plug. Plug it in halfway to the first click so you don't break the signal path—the signal still goes through the PA mixer. If you plug in all the way to the second click, the signal does not go through the PA mixer, just to the tape.

What if you want to record several instruments on one track, such as a drum mix? Assign all the drum mics to one or two output busses in the PA mixer. Plug the bus out insert jack to the tape-track input. Use two busses for stereo.

Note that recorders with RCA phono jacks are designed to accept a –10 dBV signal level. Those with XLR-type connectors handle +4 dBm, which is 12 dB higher in level than –10 dBV. The level at the insert points may not be perfectly matched for either –10 or +4. Just do the best you can in setting levels.

Monitor Mix

To monitor the quality of the signals you're recording, you generally let the PA system be your monitor system. But you may want to set up a monitor mix over headphones so you can hear what you're recording. Here are some ways to do this:

- Connect all the MDM outputs to unused line inputs on your mixer. Use those faders to set up a monitor mix. Assign them to an unused bus, and monitor that bus with headphones. If you can spare only a few inputs, plug in just one track at a time to check its sound quality.
- Use the aux 1 or aux 2 knobs to set up a monitor mix. Monitor the aux send bus over headphones. To get the most isolation while monitoring, use in-the-ear phones covered by industrial hearing protectors.

Setting Levels

Set recording levels with the PA mixer's trim or input atten knobs. This affects the levels in the PA mix, so be sure to discuss your trim adjustment in advance with the PA mixer operator. If you turn down an input trim, the PA operator must compensate by turning up that channel's fader and monitor send.

As we said, if the PA operator changes the input trim during the show, these changes will show up on your recording.

Set recording levels before the concert during the sound check (if any!). It is better to set the levels a little too low than too high, because during mixdown you can reduce noise but not distortion. A suggested starting level is -10 dBFS, which allows for surprises. Do not exceed 0 dBFS. Also, if you set the recording level conservatively, you are less likely to change the gain trims during the performance. You don't want to hassle the PA operator.

If you have a spare MDM, record a safety copy on it at the same time. This provides a backup in case one tape has a dropout.

Keep a tape log as you record, noting the counter times of tunes, sonic problems, and so on. You can refer to this log when you mix.

Feed the PA Mixer Direct Outs or Insert Jacks to a Recording Mixer

The previous method has a drawback: You have to adjust the PA mixer's trim controls, and this changes the PA mix slightly. A way around this is to connect the PA mixer direct outs or insert sends to the line inputs of a separate recording mixer. Connect the recording mixer insert sends to the MDM(s) (Figure 2–9). Set MDM recording levels with the recording mixer.

This method requires some compromises. You need more cables and another mixer. Also, the signal goes through more electronics, so it is not quite as clean as connecting straight to the PA mixer. If you use the PA mixer's direct outs, they usually are postfader, so any changes done to the PA mix will show up on your tape. Use insert sends instead, unless they

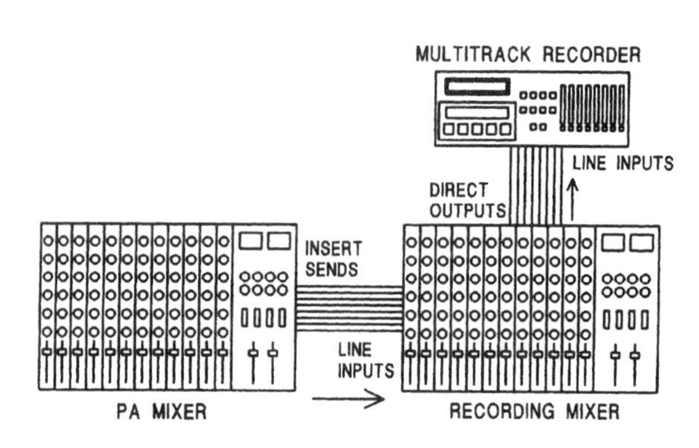

Figure 2–9
Connecting PA mixer to recording mixer to MDM.

are tied up with signal processors. Any changes the PA operator makes in the trim settings during the show will affect your recording as well.

Using the recording mixer's faders, you can set up a monitor mix. If you can hear the monitor mix well enough over headphones, you can even omit the multitrack tape recorders and attempt a live mix to a two-track recorder.

In a live mix, never turn off a mic completely unless you know for sure that it's not going to be used. Otherwise, you'll invariably miss cues. Turn down unused mics about 10 dB.

Splitting the Microphones

There's a way to have totally independent control of each microphone: Use a mic splitter. This device has one XLR-type input and two or three XLR-type outputs per mic. The signal from each mic or direct box is split three ways to feed the PA mixer, recording mixer, and monitor mixer (if used).

In the splitter, each mic or direct box plugs into an XLR that goes to a 1:1 transformer (Figure 2–10). The splitter has three feeds: one direct and two isolated. Connected directly to the mic, the direct feed goes to the mixer that supplies phantom power.

The two isolated feeds go to the other mixers. Since the transformer electrically isolates the three mixers, phantom power from one mixer can't get into the other mixers. Neither can any radio interference or cable shorts.

A high-quality transformer, such as Jensen JE-MB-C, does not degrade the signal. There may be a 1.5 dB insertion loss. Active splitters are available that let you run longer cable runs, say, 500 feet.

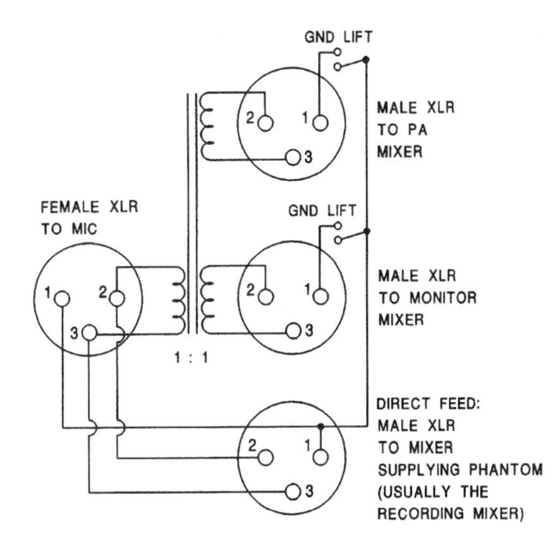

Figure 2–10
Transformer-isolated microphone splitter.

Although such a splitter is expensive, it is worth it. Isolating the signals prevents interaction and ground loops between the three mixers. You should get a clean, buzz-free signal from each mic if you use such a splitter.

Splitting has several advantages. Each mix engineer can work without interfering with the others. The FOH engineer can change trims, level, or EQ and it will have no effect on the signals going to the recording engineer. Also, a splitter provides consistent, unprocessed recordings of the mic signals. This consistency makes it easy to edit between different performances.

What's more, splitters let you use mic preamps on stage if you wish. That way, the signal path from each mic is short, which improves sound quality.

Splitters come in different sizes. A small splitter divides the signal from just one mic. A multichannel splitter divides the signals from 16, 24, or more mics.

Splitters in Use

To use a splitter, plug each mic into a splitter input. Connect one set of splitter outputs to the recording snake. Connect another set of splitter outputs to the PA snake, and so on.

The best splitters have a ground-lift switch on each channel. This switch connects or disconnects (floats) the cable shield from pin 1 of the XLR connector. When the ground-lift switches are set correctly, you should get no ground loops or their resulting buzzes.

How do you set the ground-lift switches?

1. First, turn *off* phantom power in each console.
2. Plug the snakes into their repective mixers.
3. Power up the mixers.
4. Decide which mixer you want to supply phantom power (usually the PA mixer). Connect the splitter's direct feed to that mixer's snake.
5. Make sure the direct feed's ground-lift switches (if any) are set to *ground*, not *lift*. Otherwise phantom power won't work.
6. Go to the mixer connected to the direct feed. Turn down its master faders and switch on phantom power.
7. The other mixers should already have phantom turned *off*. Only one mixer should supply phantom.
8. In the splitter, connect one isolated feed to another mixer.

9. Find the ground-lift switches for that feed, and set them to the position where you monitor the least hum and buzz (usually lifted).
10. Repeat steps 8 and 9 for the other mixer, if any.

If you follow these steps, you get several benefits:

- The mic-cable shields are tied to one mixer's ground. So any hum interference picked up on the mic cables is drained to ground.
- Phantom power needs a ground connection to work, and it gets that ground from the mixer plugged into the direct feed.
- Usually, the mics are grounded to only one mixer, and this prevents ground loops.

If all else fails, set each ground-lift switch to the position where you monitor the least hum. You might need to lift the ground on a channel used by a phantom-powered mic. If so, power the mic from a portable phantom supply.

Multitrack Recording in a Truck

Here is the ultimate setup. Each mic is split three ways to feed the snake boxes for the recording, reinforcement, and monitor consoles. A long multiconductor snake is run to a recording truck or van parked outside the concert hall or club.

In the truck, the snake connects to a mixing console, which is used to submix groups of mics and route their signals to a multitrack tape machine. Sometimes two tape machines are run in parallel to provide a backup in case one fails. Or two analog machines can be synchronized with SMPTE time code to increase the number of tracks available.

It's common to have two 24-track machines in the truck for redundancy. If the performance will take longer than the tape length, you need to stagger the recordings on both machines. That is, start the second recorder before the end of the first tape. You might start the second machine 10 minutes after starting the first one. That way, most of the performance will be on two tapes so you have a backup.

Tips on building a recording truck are given in Chapter 6.

3

BEFORE THE SESSION: PLANNING

Ready to record a live gig? The recording will go a lot smoother if you plan what you're going to do. So sit down, grab a pen, and make some lists and diagrams as described here. We go over the steps to plan a recording.

Preproduction Meeting

Call or meet with the PA company and the production company putting on the event. Find out the date of the event, location, phone numbers of everyone involved, when the job starts, when you can get into the hall, when the second set starts, and other pertinent information. Decide who will provide the split, which system will be plugged in first, second, and so on. Draw block diagrams for the audio system and communications (comm) system. Determine who will provide the comm headphones.

If you're using a mic splitter, work out the splitter feeds. The mixer getting the direct side of the split provides phantom power for condenser mics that are not powered on stage. If the house system has been in use for a long time, give the house the direct side of the split.

Overloud stage monitors can ruin a recording, so work with the sound-reinforcement people toward a workable compromise. Ask them to start with the monitors quiet because the musicians always want them turned up louder.

Make copies of the meeting notes for all participants. Don't leave things unresolved. Know who is responsible for supplying what equipment.

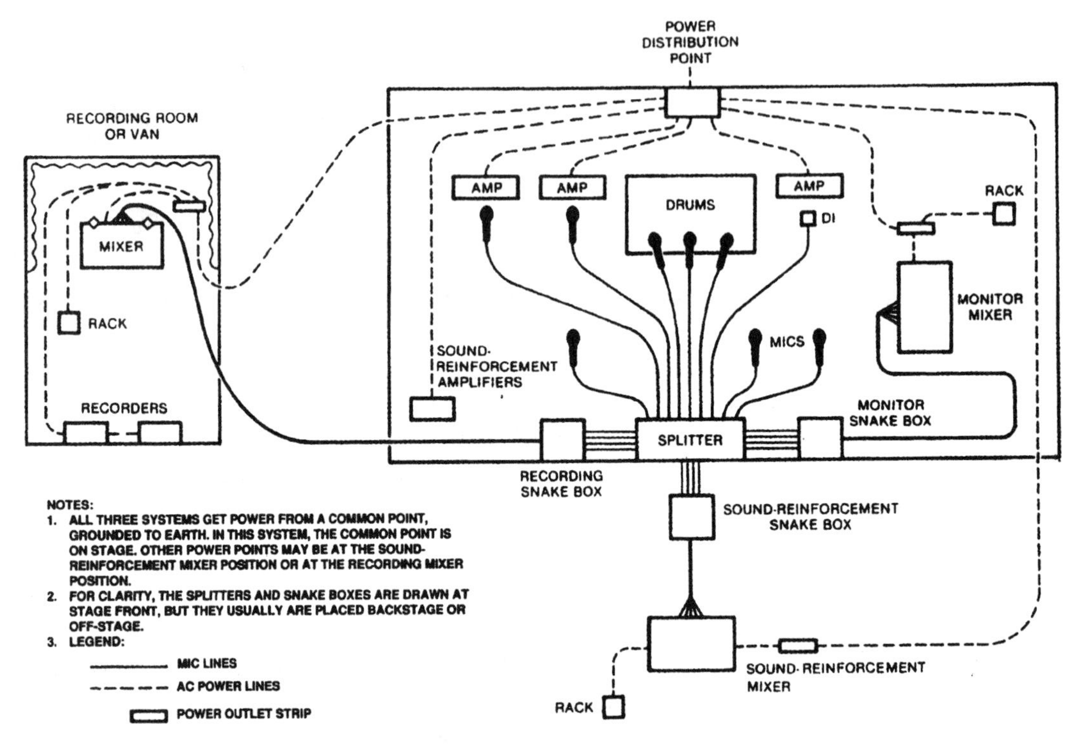

Figure 3–1 Typical layout for an on-location recording of a live concert.

Figure 3–1 shows a typical equipment layout worked out at a preproduction meeting. Three systems are in use: sound-reinforcement, recording, and monitor mixing. The mic signals are split three ways to feed these systems.

Site Survey

If possible, visit the recording site in advance and go through the following checklist:

- Check the AC power to make sure the voltage is adequate, the third pin is grounded, and the waveform is clean.
- Listen for ambient noises—ice machines, coolers, 400-Hz generators, heating pipes, air conditioning, nearby discos, and the like. Try to have these noise sources under control by the day of the concert.
- Sketch the dimensions of all rooms related to the job. Estimate distances for cable runs.
- Turn on the sound-reinforcement system to see if it functions okay by itself (no hum and so on). Turn the lighting on at various levels with

the sound system on. Listen for buzzes. Try to correct any problem so that you don't document bad PA sound on your tape.

- Determine locations for any audience or ambience mics. Keep them away from air-conditioning ducts and noisy machinery.
- Plan your cable runs from the stage to the recording mixer.
- If you plan to hang mic cables, feel the supports for vibration. You may need microphone shock mounts. If there's a breeze in the room, plan on taking windscreens.
- Make a file on each recording venue including the dimensions and the location of the circuit breakers.
- Determine where the control room will be. Find out what surrounds it—any noisy machinery?
- Visit the site when a crowd is there to see where there may be traffic problems.

Mic List

Now write down all the instruments and vocals in the band. If you want to put several mics on the drum kit, list each drum that you want to mike. As for keyboards, decide whether you want to record off each keyboard's output or off the keyboard mixer (if any).

Next, write down the mic or direct box you want to use on each instrument; for example,

1	Bass	DI
2	Kick	AKG-D112
3	Snare	Shure Beta 57
4	Hi hat	Crown CM-700
5	Small rack tom	57
6	Big rack tom	57
7	Small floor tom	Sennheiser MD-421
8	Big floor tom	421
9	Cymbals overhead left	Shure SM81
10	Cymbals overhead right	SM81
11	Lead guitar	57
12	Rhythm guitar	57
13	Keyboard mixer	DI
14	Lead vocal	Beyer M88
15	Harmony vocal	Crown CM-311A

Make two copies of this mic list. At the gig, place one list by the stage box and the other by the PA mixer. Each list will act as a guide to keep things organized.

Track Sheet

Next, decide what will go on each track of your multitrack tape recorder. If you have enough tracks, your job is easy: Just assign each instrument or vocal to its own track. Bass to track 1, kick to track 2, and so on.

What if you have more instruments than tracks? Suppose you have an eight-track recorder, but 15 instruments and vocals (including each part of the drum set). You'll need to assign several instruments or vocals to the same track. That is, you will set up a submix.

Let's say the drum kit includes a snare, kick drum, two rack toms, two floor toms, hi-hat, and cymbals. If you want to mike everything individually, that's nine mics including two for the cymbals. But you don't need to use nine tracks. Assign or group those mics to buses 1 and 2 to create a stereo drum mix. Connect buses 1 and 2 to tape tracks 1 and 2. At the sound check, set up a submix of all the drum mics, and assign them to buses 1 and 2. You control the overall level of the drum mix with submaster faders 1 and 2 (also called bus faders or group faders).

Use tracks 3 through 8 for amps and vocals (as in the following example). Feed tracks 3–8 from insert sends.

Track/instrument

1. Drum mix L.
2. Drum mix R.
3. Bass.
4. Lead guitar.
5. Rhythm guitar.
6. Keys mix.
7. Lead vocal.
8. Harmony vocal.

Block Diagram

Now that your track assignments are planned, you can figure out what equipment you'll need. Draw a block diagram of your recording setup from input to output (Figure 3–2). Include mics, mic stands, DI boxes, cables, snake, mixer, multitrack tape recorder(s), and powering. On your diagram, label the cable connectors on each end so you'll know what kind

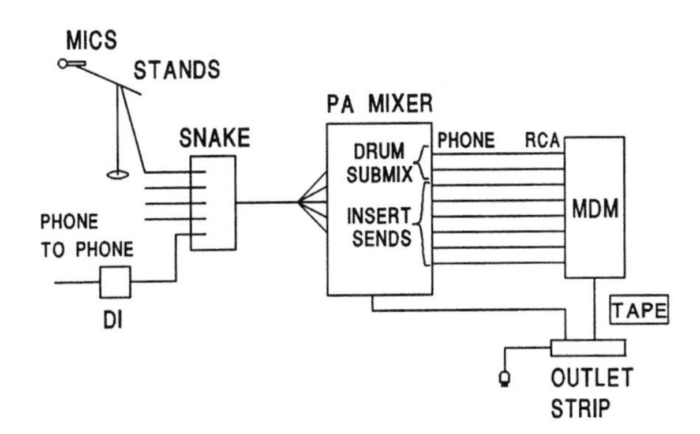

Figure 3–2
Example block diagram of recording setup.

of cables to bring. It is a good idea to keep a file of system block diagrams for various recording venues.

In Figure 3–2, the block diagram shows a typical recording method: feeding PA-console insert jacks to a modular digital multitrack. We use this example throughout the rest of the chapter.

Equipment List

From your block diagram, generate a list of recording equipment. Based on Figure 3–2, you would need the following PA and recording gear (not including mics, power amps, and speakers):

- Two guitar cables.
- Mic cables.
- Mic stands and booms.
- Snake.
- PA mixer and effects rack.
- Eight male-phone to male-RCA cables (this can be a snake).
- Multitrack recorder.
- Outlet strip.
- Extension cord.
- Recording tapes (bring enough for the duration of the gig).

Don't forget the incidentals: a cleaning tape or alcohol and cotton swab, pen, notebook, flashlight, guitar picks, heavy-duty guitar cords, drum keys, spare tape, mic pop filters, gaffer's tape, tuner, ear plugs, audio-connector adapters, electrical three-to-two AC adapters, ground-lift adapters, in-line pads, in-line polarity reversers, spare cables, gooseneck lights for the console, spare batteries—and aspirin.

Bring a tool kit with screwdrivers, pliers, soldering iron and solder, AC-outlet checkers, fuses, a pocket radio to listen for interference, ferrite beads of various sizes for RFI suppression, canned air to shoot out dirt, cotton swabs and pipe cleaners, and De-Oxit from Caig Labs to remove oxide from connectors.

Check off each item on the list as you pack it. After the gig, you can check the list to see whether you reclaimed all your gear.

Preparing for Easier Setup

You want to make your setup as fast and easy as possible. Here are some tips to help this process.

Put It on Wheels

Mount your console and recorders in protective carrying cases. Install casters or swivel wheels under racks and carrying cases so you can roll them in. Rolling is so much easier than lifting and carrying.

You might permanently install the MDMs in SKB carrying cases that act as racks. When a remote job comes up, just grab them and go.

A very helpful item is a dolly or wheeled cart to transport heavy equipment into the venue. Consider getting some lightweight tubular carts. Being collapsible, they store easily in your car or truck.

One maker of equipment carts is Rock 'n' Roller, who advertises in the *Musician's Friend* catalog, www.musiciansfriend.com. Another cart is the Remin Kart-a-Bag, www.kart-a-bag.com.

Pack mics, headphones, and other small pieces in trunks or milk crates. Caution: Keep tapes separated from magnets, such as in headphones, monitor speakers, and dynamic mics. (However, tape erasure is much less of a problem with digital media like DAT than with analog tapes.)

You might want to build a mic container: a big box full of foam rubber with cutouts for all the mics. Or construct a wheeled cabinet with drawers for mics, DIs, and speaker cables.

In an article in the September–October 1985 issue of *db* magazine, remote recording engineer Ron Streicher offers these suggestions:

> Especially for international travel, make sure your documentation is up to date and matches the equipment you're carrying. Make a list of everything you take: all the details, such as each pencil, razor blade, connector, etc.
>
> Also make sure your insurance is up to date. You need insurance for en route as well as at the destination.

I organize my cases so I know where every item is. They're ready to go anytime and make setup much faster. The cables are packed with their associated equipment, not in a cables case. I check everything coming and going, and try to have 100 percent redundancy, such as a small mixer to substitute for the large console.

Mic Mounts

If you'll be recording a singer/guitarist, take a short mic mount that clamps onto the singer's mic stand. Put the guitar mic in the mount. Also bring some short mounts to clamp onto drum rims and guitar amps. By using these mounts, you eliminate the weight and clutter of several mic stands.

Some examples of short mounts are the Mic-Eze units by Ac-cetera. They have standard $\frac{5}{8}$-inch 27 thread and mic clamps that either spring shut or are screw tightened. Flex-Eze is two clamps joined by a short gooseneck. Max-Eze is two clamps joined by a rod; Min-Eze is two clamps joined by a swivel. Ac-cetera, Inc. is at 3120 Banksville Road, Pittsburgh, PA 15216, phone 800–537–3491, e-mail aaps@pgh.nauticom.net.

Snakes and Cables

You can store mic cables on a cable spool, available in the electrical department of a hardware store. Wrap one mic cable around the spool, plug it into the next cable and wrap it, and so on. No more tangled cables.

A snake can be wrapped around a garden-hose spool or two widely spaced vertical posts screwed inside the bottom of a trunk (see Figure 3–3).

Figure 3–4 shows a custom snake spool with a center partition. Wind a few feet of the XLR/pigtail end around the left half. Wind the rest of the snake around the right half. Carry the snake up to the stage and put down the stage box. Then carry the spool back to the console while unwinding the snake. Lay the spool on its side by the console, then unwind the XLR end and plug it into the console.

Commercial snake reels are made by such companies as Whirlwind, ProCo, and Hanny (www.hannay.com).

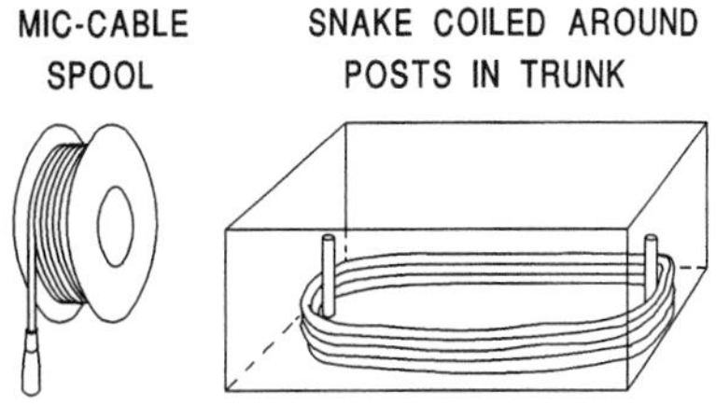

MIC-CABLE SPOOL

SNAKE COILED AROUND POSTS IN TRUNK

Figure 3–3
Some cable-storage methods.

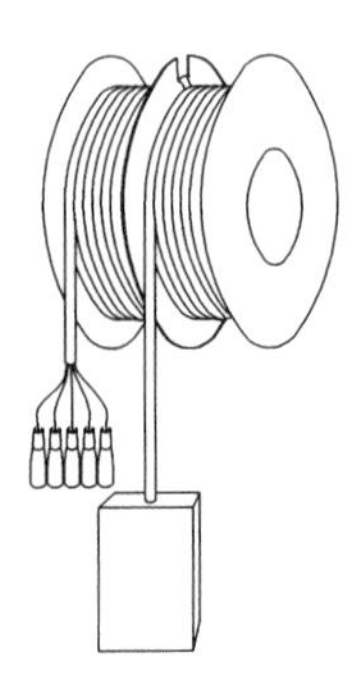

Figure 3–4
A two-section snake spool.

Use wire ties to join cables that you normally run together, such as PA sends and returns.

Snake hookup is quicker if the snake has a multipin twist-lock connector (such as Whirlwind W1 or W2). This connector plugs into a mating connector that divides into several male XLRs. Those XLRs plug into the mixing console. Leave the XLRs in the console carrying case. You'll find that the snake is easier to handle without the XLR pigtails.

For a clean, rapid hookup of drum mics, put a small snake near the drum kit, and run it to the main stage box. Snakes are made by such companies as Whirlwind, ProCo, and Horizon.

Check that your mic cables are wired in the same polarity—pin 2 hot.

You might use three-conductor shielded mic cables (hot, cold and ground leads in a shield). Connect the shield to ground only at the male XLR end. Also use cables with 100 percent shielding. Those measures enhance the protecting capability of the shield and reduce pickup of lighting buzzes.

In XLR-type cable connectors, do not connect pin 1 to the shell, or you may get ground loops when the shell contacts a metallic surface.

Label all your cables on both ends according to what they plug into; for example, DSP-9 effects in, track 12 out, power amp in, snake line out. Or you might prefer to number the cables near their connectors. Cover these labels with clear heat-shrink tubing.

Label both ends of each mic cable with the cable length. Put a drop of glue on each connector screw to temporarily lock it in place.

Rack Wiring

You can speed up the console wiring by using a small snake between the rack and the console and between the multitrack tape recorders and the console. When packing, plug the snakes into the rack gear and multitracks, then coil the snake inside the rack and the multitrack carrying case. In other words, have all your equipment prewired. At the gig, pull out a bundled harness and plug it into the console jacks.

You might be feeding your multitrack from the insert jacks of a PA company's console. If so, use a snake with tip/ring/sleeve stereo phone plugs at the console end. Carry some TRS-to-dual-mono phone adapters to handle consoles that have separate jacks for insert send and receive.

Some engineers prefer to make a clearly marked interface panel on the rear of the racks, and plug into the panel. This is easier than trying to find the right connectors on each piece of equipment.

Small bands might get by with all their equipment in a single, tall rack. Mount a small mixer on top, wired to effects in the middle, with a coiled snake and power cord on the bottom.

Small snakes for rack and multitrack connections are made by Hosa, Horizon, and ProCo, among others.

Other Tips

Here are some more helpful hints for successful on-location recordings:

- Plan to use a talkback mic from the board to the stage monitors during sound checks. You might bring a small instrument amp for talkback so that you can always be heard.
- Hook up and use unfamiliar equipment before going on the road. Don't experiment on the job.
- Consider recording with redundant (double) systems so you have a backup if one fails.
- If a concert will be longer than the running time of a reel of tape, switch reels at intermissions. Another method is to feed two identical tape machines the same signal in parallel. Record on one machine. As the reel of tape nears the end, start recording on the second machine so that none of the performance is lost. Edit the two tapes together back in the studio.
- Walkie-talkies are okay for preshow use, but don't use them during the performance because they cause RF interference. Use hard-wired communications headsets. Assistants can relay messages to and from the stage crew while you're mixing.
- During short set changes, use a closed-circuit TV system and light table (or fax machines) to show what set changes and mic-layout changes are coming up next; transmit this information to the monitor mixer and sound-reinforcement mixer.

- Don't put tapes through airport X-ray machines because the transformer in these machines is not always well shielded. Have the tapes inspected by hand.
- Hand carry your mics on airplanes. Arrange to load and unload your own freight containers, rather than trusting them to airline freight loaders. Expect delays here and at security checkpoints.
- Get a public-liability insurance policy to protect yourself against lawsuits.
- Call the venue and ask directions to the load-in door. Make sure that someone will be there at setup time to let you in. Ask the custodian not to lock the circuit-breaker box the day of the recording.
- A few days before the session, check out the parking situation.
- Just before you go, check out all your equipment to make sure it's working.
- Arrive several hours ahead of time for parking and setup. Expect failures—something always goes wrong, something unexpected. Allow 50 percent more time for troubleshooting than you think you'll need. Have backup plans if equipment fails.
- In general, plan everything in advance so you can relax at the gig and have fun.

By following these suggestions, you should improve your efficiency—and your recordings—at on-location sessions.

Some of the information this chapter was derived from two workshops presented at the 79th convention of the Audio Engineering Society in October 1985. These workshops were On the Repeal of Murphy's Law—Interfacing Problem Solving, Planning, and General Efficiency On-Location, given by Paul Blakemore, Neil Muncy, and Skip Pizzi; and Popular Music Recording Techniques, given by Paul Blakemore, Dave Moulton, Neil Muncy, Skip Pizzi, and Curt Wittig.

4

AT THE SESSION: SETUP AND RECORDING

Okay! You've arrived at the venue. After parking, offload your gear to a holding area, rather than on stage, because gear on stage likely will need to be moved.

Learn the names of the PA company crew members and be friendly. These people can be your assets or your enemies. Think before you comment to them. Try to remain in the background and do not interfere with their normal way of doing things (for example, take the secondary side of the split).

Power and Grounding Practice

At the job, you need to take special precautions with power distribution, interconnecting multiple sound systems, and electric guitar grounding.

Power Distribution System

Consider buying, renting, or making your own single-phase power distribution system (distro). It will greatly reduce ground loops and increase reliability. Figure 4-1 shows a suggested AC power distribution system. The amp rating of the distro's main breaker box should exceed the current drain of all the equipment that will be plugged into the distro system.

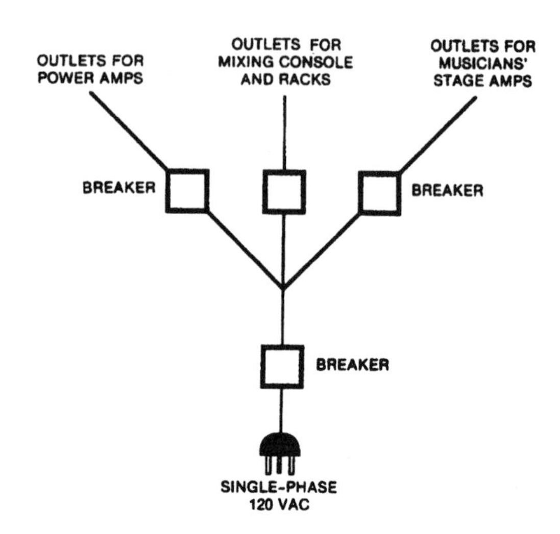

Figure 4–1
An AC power distribution system for a touring sound system.

Furman makes a model ACD-100 AC power distro. It distributes a 100-amp feed (which you get from a breaker box) to five 20-amp, 120-V circuits. The unit works on 120-V, 240-V, or 208-V three-phase circuits. Furman's web site is at http://www.furmansound.com.

Power Source

If you're using a remote truck, find a source of power that can handle the truck's power requirements, usually at a breaker panel. Some newer clubs have separate breaker boxes for sound, lights, and a remote truck. Find out whether you'll need a union electrician to make those connections. Label your breakers.

Check that your AC power source is not shared with lighting dimmers or heavy machinery; these devices can cause noises or buzzes in the audio.

The industry-standard power connector for high-current applications is the Cam-lok, a large cylindrical connector. Male and female Cam-loks join together and lock when you twist the connector ring. Distro systems and power cables with Cam-lok connectors can be rented from rental houses for film, lighting, electrical equipment, or entertainment equipment. One such rental house is Mole-Richardson, 213-851-0111, http://www.mole.com.

Use an adapter from Cam-lok to bare wires. Pull the panel off the breaker box, insert the bare wires, and connect the Cam-lok to your truck's power. CAUTION: Have an electrician do the wiring if you don't know what you're doing. A union electrician might be required anyway. Some breaker boxes have Cam-loks already built in.

To reduce ground-loop problems, get on the same power that the PA is using. From that point, run your distribution system or at least run one or two thick (14 or 16 gauge) extension cords to your recording system. These cords may need to be 100–200 feet long. Plug AC outlet strips into the extension cord, then plug all your equipment into the outlet strips.

If your recording system consists of one or two multitrack tape recorders that will connect to the PA mixer, simply plug into the same outlet strip that the PA is using.

Measure the AC line voltage. If the AC voltage varies widely, use a line voltage regulator (power conditioner) for your recording equipment. If the AC power is noisy, you might need a power isolation transformer.

Check AC power on stage with a circuit checker. Are grounded outlets actually grounded? Is there low resistance to ground? Are the outlets of the correct polarity? There should be a substantial voltage between hot and ground, and no voltage between neutral and ground.

Some recording companies have a gasoline-powered generator ready to switch to if the house power fails. If there are a lot of lighting and dimmer racks at the gig, you might want to put the truck on a generator to keep it isolated from the lighting power.

Interconnecting Multiple Sound Systems

If you encounter an unknown system where balanced audio cables are grounded at both ends, you might want to use some cable ground-lift adapters (Figure 4–2) to float (remove) the extra pin-1 ground connection at equipment inputs.

If you hear a hum or buzz when the systems are connected, first make sure that the signal source is clean. You might be hearing a broken snake shield or an unused bass-guitar input. If the hum persists, experiment with flipping the ground-lift switches on the splitter and on the direct boxes. On some jobs, you need to lift almost every ground; on others, you need to tie

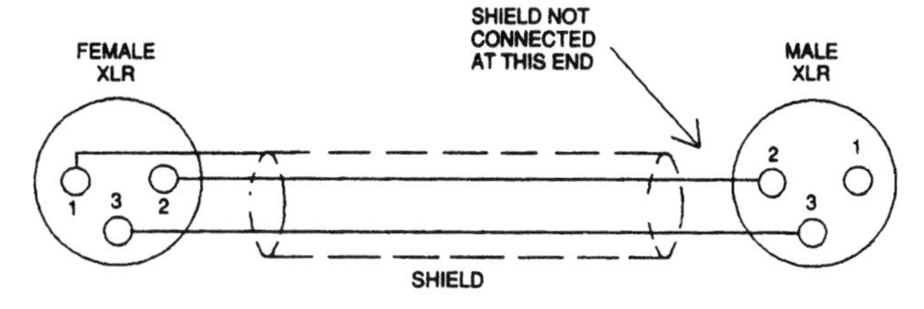

Figure 4–2
A ground-lift adapter for balanced line-level cables.

all the grounds. The correct ground-lift setting can change from day to day due to a change in the lighting. Expect to make some trial and error adjustments.

If the PA has serious hum and buzz problems, offer help. You can hear buzzes in your quiet truck that they can't hear over the main system with noise in the background.

Often, a radio station or video crew will take an audio feed from your mixing console. In this case, you can prevent a hum problem by using a console with transformer-isolated inputs and outputs. Or you can use a 1:1 audio isolation transformer between the console and the feeds. For best isolation, use a distribution amp with several transformer-isolated feeds. Lift the cable shield at the input of the system you're feeding. Some excellent isolation transformers are made by Jensen (213-876-0059, http://www.jensen-transformers.com).

John Fasarano, owner of Studio on Wheels in Glendale, California, describes how he prevents ground loops and hum: "I usually don't connect ground on the AC power. This solves 80 percent of the hum problems. I also flip the ground lifts or run AC into the house, and have them plug into that."

Guy Charbonneau, owner of Le Mobile, Hollywood, has this to say:

> The truck uses 3-phase, filtered 240V power with a 25 kVA transformer having six different taps. I don't use the neutral; it carries a lot of current from lighting systems. The truck chassis is grounded. I use no ground lifts 99 percent of the time; I carry through the shield with the sound company. In some clubs, I bring 50 amps to the stage on a 220V line with a distribution box. All the musicians' instruments and the club console plug in there. This prevents ground loops and AC line noises from coffee machines and dishwashers. When working with a big P.A. company, I just ask for their split. Often a Y-adapter works. (Source: "AES Workshop: The State of the Art of Remote Recording," *db* magazine, September–October 1989.)

Connections

After unpacking, place one mic list by the stage box so you know what to plug in where. Place a duplicate list by the PA mixer. Attach a strip of white tape just below the mixer faders. Use this strip to write down the instrument that each fader affects.

Based on the mic list you wrote, you might plug the bass DI into snake input 1, plug the kick mic into snake input 2, and so on. Label fader

1 "bass," label fader 2 "kick," and so forth. Also plug in equipment cables according to your block diagram.

Have an extra microphone and cable offstage ready to use if a mic fails.

Don't unplug mics plugged into phantom power, because this will make a popping noise in the sound-reinforcement system.

Running Cables

To reduce hum pickup and ground-loop problems associated with cable connectors, try to use a single mic cable between each mic and its snake-box connector.

Avoid bundling together mic cables, line-level cables, and power cables. If you must cross mic cables and power cables, do so at right angles and space them vertically.

Plug each mic cable into the stage box, then run the cable out to the each mic and plug it in. This leaves less of a mess at the stage box. Leave the excess cable at each mic stand so you can move the mics. Don't tape down the mic cables until the musicians are settled.

It is important that audience members do not trip over your cables. In high-traffic areas, cover cables with rubber floor mats or cable crossovers (metal ramps). At least tape them down with gaffer's tape.

It helps to set up a closed-circuit TV camera and TV monitor to see what's happening on stage. You need to know when mics get moved accidentally, when singers use the wrong mic, and so on.

Recording-Console Setup

Here is a suggested procedure for setting up the recording system efficiently:

1. If the console is set up in a dressing room or locker room, add some acoustic absorption to deaden the room reflections. You might bring a carpet for the floor plus fiberglass insulation and packing blankets for the walls.
2. Turn up the recording monitor system and verify that it is clean.
3. Plug in one mic at a time and monitor it to check for hums and buzzes. Troubleshooting is easier if you listen to each mic as you connect it, rather than plugging them all in and trying to find a hum or buzz.

4. Check and clean up one system at a time: first the sound-reinforcement system, then the stage-monitor system, then the recording system. Again, this makes troubleshooting easier because you have only one system to troubleshoot.
5. Use as many designation strips as you need for complex consoles. Label the input faders bottom and top. Also label the monitor-mix knobs and the meters.
6. Monitor the reverb returns (if any) and check for a clean signal.
7. Make a short test recording and listen to the playback.
8. Verify that left and right channels are correct and that the pan-pot action is not reversed audibly.
9. If you are setting up a separate recording monitor mix, do a preliminary pan-pot setup. Panning similar instruments to different locations helps you identify them.

Mic Techniques

Usually the miking is left up to the PA company. But you should know about some mic-related problems, such as feedback, leakage, room acoustics, and noise. Here are some ways to control these problems:

- Use directional microphones, such as cardioid, supercardioid, or hypercardioid. These mics pick up less feedback, leakage, and noise than omnidirectional mics at the same miking distance.
- With vocal mics, aim the null of the polar pattern at the floor monitors. The null (area of least pickup) of a cardioid is at the rear of the mic: 180° off axis. The null of a supercardioid is 125° off axis; hypercardioid is 110°.
- Mike close. Place each mic within a few inches of its instrument. Ask vocalists to sing with lips touching the mic's foam pop filter.
- Use direct boxes. Bass guitar and electric guitar can be recorded direct to eliminate leakage and noise in their signals. However, you might prefer the sound of a miked guitar amp. You could record the guitar direct from its effects boxes, then use a guitar-amp emulator during mixdown. Note that sequencers and some keyboards have high-level outputs, so their DI boxes need transformers that can handle line level.
- Use contact pickups. On acoustic guitar, acoustic bass, and violin, you can avoid leakage by using a contact pickup. Such a pickup is sensi-

tive to only the instrument's vibration, not so much to sound waves. The sound of a pickup is not as natural as a microphone, but a pickup may be your only choice. Consider using both a pickup and a microphone on the instrument. Feed the pickup to the PA and monitor speakers, and feed the mic to the recording mixer.

To reduce breath pops with vocal mics, be sure to use foam pop filters. Allow a little spacing between the pop filter and the mic grille. It also helps to switch in a low-cut filter (100 Hz high-pass filter).

When you're recording a band that has been on tour, should you use its PA mics or your own mics? In general, go with its mics. The artists and PA company have been using the band's mics for a while and may not want to change anything. Most mics currently used in PA are good quality anyway, unless they are dirty or defective.

If you're not happy with their choice, you could add your own instrument mics. Let the PA people listen to the sound in the recording truck or in headphones. If it sounds bad because of the mic choice, ask, "Would it be okay if we tried a different mic (or mic placement)?" Usually it's all right with them—it's a team effort.

Electric Guitar Grounding

While setting up mics, you need to be aware of a safety issue with the electric guitar. Electric guitar players can receive a shock when they touch their guitar and a mic simultaneously. This occurs when the guitar amp is plugged into an electrical outlet on stage, and the mixing console (to which the mics are grounded) is plugged into a separate outlet across the room. If you're not using a power distro, these two power points may be at widely different ground voltages. So a current can flow between the grounded mic housing and the grounded guitar strings.

Caution: Electric guitar shock is especially dangerous when the guitar amp and the console are on different phases of the AC mains.

It helps to power all instrument amps and audio gear from the same AC distribution outlets. If you lack a power distro, run a heavy extension cord from a stage outlet back to the mixing console (or vice versa). Plug all the power-cord ground pins into grounded outlets. That way, you prevent shocks and hum at the same time.

If you're picking up the electric guitar directly, use a transformer-isolated direct box and set the ground-lift switch to the minimum-hum position.

Using a neon tester or voltmeter, measure the voltage between the electric guitar strings and the metal grille of the microphones. If there is a voltage, flip the polarity switch on the amp. Use foam windscreens for additional protection against shocks.

Audience Microphones

If you have enough mic inputs, you can use two audience mics to pick up the room acoustics and audience sounds. This helps the recording sound "live." Without audience mics, the recording may sound too dry, as if it were done in a studio.

One easy technique is to tape two boundary mics about 4 feet apart on the front face of the stage (Figure 4–3). You don't have to hang any microphones, but the mics may pick up conversation in the first few rows of seating. You might try a pair of PZMs on the side walls.

Another method is to aim two cardioid mics at the audience. Hang them high over the front row of the audience aiming at the back row (Figure 4–3). You might be able to set up a stereo pair on a mic stand and tape off the seats around the stand so that it doesn't get bumped.

If the audience mics are far back in the hall—100 feet from the stage, or at FOH, for example—they pick up the band's sound with a delay. When mixed with the close mics, the audience mics add an echo. Prevent this by placing the audience mics fairly near the stage. Or, during mix-down, delay the stage-mics' signals so that they coincide with the audience-mics' signals.

Here is one way to create this delay: Record the stage-mic mix on two tracks of a digital editor. Record the audience mics on two other tracks. Slide the stage-mic tracks to the right in time (that is, delay them) so that they coincide with the audience-mic tracks.

What if you lack enough tracks for the audience mics? Record them on a two-track DAT. Load this recording into your digital editor along with the stage-mic mix. Align the two recordings in time as just described.

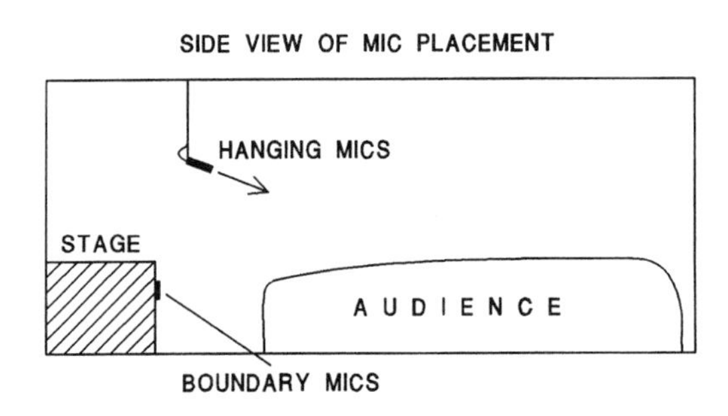

Figure 4–3
Some audience miking techniques.

If the audience mics are run through the PA mixer, leave the audience mics unassigned in that mixer to prevent feedback.

To get more isolation from the house speakers in the audience mics, use several mics close to the audience. Some engineers put up 4 audience mics maximum; some use 8 to 10. Use directional mics and aim the rear null at the PA speakers.

Another option is to *not* mike the audience or not use the audience tracks. Instead, during mixdown, simulate an audience with audience-reaction CDs. Simulate room reverb with an effects unit.

Setting Levels and Submixes

Now that the mics are set up, you might have time for a sound check. That's when you set recording levels. Have the band play a loud song. Locate a mixer input module that is directly feeding a tape track. Set the input trim (mic preamp gain) to get the desired recording level on each track. On an MDM recorder, you might set each track's level to peak around –10 dB, which allows for surprises if someone plays louder during the live gig.

Most of the mixer channels feed tape tracks directly from the insert sends. On those channels, any fader moved during the gig will not affect the levels going on tape. Why? In most mixers, the insert send is prefader. That is, the signal at the insert-send jack is not affected by the fader. However, you may encounter PA consoles where some insert sends are tied up with signal processors. You must use those channels' direct-out jacks instead, which usually are postfader (unless they can be switched to prefader).

Now let's set up the drum submix. (Ideally, you would do this with the PA turned down, and monitor over headphones or Nearfield monitors in a separate room.) Assign each drum mic to busses 1 and 2, and pan each mic as desired. Put the faders for busses 1 and 2 at design center—the shaded area about $\frac{1}{2}$ to $\frac{3}{4}$ up. Set each drum-mic fader to about –10. Set a rough drum-kit mix with the input trims while keeping the mixer meters around 0. Fine-tune the drum mix with the faders, and set the recording level with the drum-mix bus faders.

Here is another way to create the drum submix. Have the drummer hit each drum repeatedly, one at a time, as you adjust the input trims to prevent clipping. For example, ask the drummer to bang on the kick drum. Turn down the kick-drum's input trim all the way. Slowly bring it up until the clip LED (overload light) flashes. Then turn down the input trim about 10 dB to allow some headroom.

When all the drum trims are set, set a drum mix with the faders and set the recording level with the bus 1 and 2 faders. Caution: Any changes to the drum mix or drum-mix level will show up on your tape.

Recording

If your recording will be synched later with a videotape by using SMPTE time code, record the video time-code feed on a spare tape track.

A few minutes before the band starts playing, roll the tape. Keep a close eye on recording levels. If a track is going into the red, slowly turn down its input trim and note the tape-counter time where this change occurred.

Caution: If you are recording off the PA mixer, turning down its input trim will affect the PA levels. The PA operator will need to turn up the corresponding monitor send and channel fader.

This is a touchy situation that demands cooperation. Ideally, you set enough headroom during the sound check so you won't have to change levels. But be sure the PA operator knows in advance that you might need to make changes. Ask the operator whether he or she wants to adjust the gain trims for you, so the operator can adjust corresponding levels at the same time. Thank the operator for helping you get a good recording.

If you are recording with a splitter and mic preamps on stage, assign someone to watch the levels and adjust them during the concert. Preamps with meters allow more precise level setting than preamps with clip LEDs.

Keep a track sheet and tape log as you record. For each song in the set list, note the tape-counter time when the song starts. Later, during mixdown, you can fast-wind to those counter times to find songs you want to mix. Also note where any level changes occurred so you can compensate during mixdown. It helps to note a counter time when the signal level was very high. When you mix the tape you can start at that point in setting your overall mix levels.

Teardown

After the gig, pack away your mics first because they may be stolen or damaged. Refer to your equipment list as you repack everything. Note equipment failures and fix broken equipment as soon as possible.

After you haul your gear back to the studio, it is time for mixing and editing, covered next.

5

AFTER THE SESSION: MIXING AND EDITING

At this point, you brought your multitrack tapes back to the studio, ready to mix. First, listen to each track by itself and erase anything you don't want to keep. For example, erase some talking on the vocal track during the song intro. Or erase a thump on a guitar track that happened when someone bumped into the mic stand.

Overdubs

If the band members are available, they might want to overdub parts of tracks to correct mistakes. Be careful to match the sound of the instrument to the sound of the track.

For example, I did a 16-track live recording on two MDMs. All the vocal mics were mixed to one track. The producer wanted to rerecord all the vocals in the studio to get a better mix and a better performance. We rented an MDM, synced it with the other two, and overdubbed studio vocals on the rented unit.

Mixdown

Mixdown procedures can be found in standard recording textbooks, so they will not be covered here. However, you need to consider some things when mixing recordings made on location.

Live gig tapes are long, nonstop programs. It's hard to mix continuously for an hour or two without making a mistake. A better procedure is to set up a mix for the first song, record it on DAT, set up a mix for the next song, record it on DAT, and so on. Then edit the mixes together to sound like a single concert.

When mixing each song to DAT, be sure to include several seconds of crowd noise or applause before and after each song. You will use this crowd sound when editing later.

Keep the audience mics in the mix to make it sound live. Audience mics can muddy the sound if mixed in too loudly. Keep them down in level, just enough to add some atmosphere. Bring them up gently to emphasize crowd reactions. You can reduce muddiness by rolling off the lows in the audience mics.

As stated in Chapter 4, the audience mics might sound delayed compared to the stage mics, causing an echo. Chapter 4 suggested some ways to remove this delay.

Mixing for Surround Sound

If you are mixing a live gig for surround sound, you probably want to put the listener inside a simulated concert hall. To do this, feed the audience mics to the rear surround-sound channels. Also feed some or all the reverberation to the rear, whether it was miked or made with an effects unit. (If you need details on surround sound, see Chapter 13.)

To create the subwoofer channel, turn up an aux send equally on each input channel that has low-frequency content. Low-pass filter the aux-mix output at 120 Hz. That filtered aux becomes your LFE channel. You also could use a spare bus for the sub channel and assign bassy channels to that bus.

Panning in surround can be done with a TMH panner (about $500/channel) or a Circle Surround 5.2.5 panner/controller from RSP Technologies. Some consoles come with surround pan pots built in. Also available is MX51 software from Minnetonka Audio Software Inc., which works with the Digital Audio Labs V8 system.

To adjust the overall monitor level, you might need a six-gang pot. If you have an automated console, you could control six input modules (used only for monitoring) with a single group fader. Set the monitor level at 85 dB SPL, C-weighted, slow-reading scale and leave it there.

In stereo mixdowns, it is common to pan the bass, kick drum, snare, and lead vocal to center. Feeding all these sounds to the center speaker can cause it to overload. You might want to feed these sounds partly to front left and front right; this takes some of the load off the center speaker.

If you start with an instrument in the center speaker, you can send some of its signal to the surrounds to move the instrument toward you.

A center speaker has more output around 2 kHz than a center phantom image. The 2 kHz bump can sound annoying on vocals, so be careful with EQ.

Instruments are more distinct in surround-sound mixes. It is easier to hear what each is doing. So you may find that you need less level adjusting and less compression in surround sound than in stereo.

You might start with a standard stereo mix on two speakers, then add the center, surrounds, and sub. Adjust the mix to take advantage of the other speakers.

Percussive sounds can be distracting in the rear, but synth pads and harmony vocals work well there.

Be sure to check your surround-sound mix in stereo and mono to make sure it is compatible. The reverb level in stereo listening should match the reverb level in multichannel listening. That is, when you fold down or collapse the monitoring from 5.1 to stereo, the direct/reverb ratio should stay the same.

The TASCAM DA-88 modular digital multitrack (MDM) is a popular storage medium for discrete 5.1 mixes. A suggested track layout is

1. Left front.
2. Right front.
3. Center.
4. Low-frequency effects (LFE).
5. Left surround.
6. Right surround.

You can use track 8 for SMPTE time code. Or, send the six channels to a Prism MR-2024T converter, which creates six channels of 20-bit data across eight tracks. Tracks 7 and 8 hold 4-bit data produced by the converter.

Editing a Gig Demo Tape

At this point, you have a DAT tape of the mixes. Now you want to edit that tape to make a tight presentation: Remove long pauses between songs, edit out songs that weren't played well, and so on.

Before you can edit a DAT tape, you need a computer with a sound card and some editing software. This system is called a *digital audio workstation* (DAW).

Recording textbooks explain the basics of how DAWs work. Here, we give tips on editing your recording with a DAW. Let's run through the process step by step.

Copy Your DAT to Your Hard Disk

First, plug your DAT recorder into your sound card. Play the DAT tape of your gig, and record it onto the hard disk.

If you want the edited gig tape to end up on a CD, you should have recorded the DAT at the 44.1 K sampling rate, the same rate as CDs. If not, your DAW editing program might be able to convert from 48 K to 44.1 K. Some programs take several minutes to calculate this conversion; others can do it instantaneously.

If your editing program does not do sample-rate conversion, connect the analog output of your DAT to the analog input of your sound card (if any). Set the sampling rate in the DAW to 44.1 K.

Next, the waveform of the audio program appears on your monitor screen (Figure 5–1). You can zoom out to see the entire program or zoom in to work on tiny spans of time.

Define Songs

In the procedures that follow, use your mouse to mark the beginning and end points of music and talking that you want to keep. Please refer to Figure 5–2. Here are the steps:

1. There should be some crowd noise or applause before the first song. Place the cursor about 5 seconds before the first song starts and mark that as the beginning point. That way, you can start the edited recording by fading up on crowd noise. This helps to establish the recording as being "live." If there is no crowd noise before the first song, find some crowd noise elsewhere in your tape. Mark off a section of it about 5 seconds long and save it as a region called *Crowd.*
2. If you want this to be a gig demo tape, you might want to play only about 60 seconds of each song. That way, the listener can quickly sample all the band's styles without having to hear each song from start to finish. The edited version would be something like this:

    Song 1 intro, verse, chorus, and fade out after the chorus.
    Song 2 intro, verse, chorus, and fade out after the chorus.
    Song 3 intro, verse, chorus, and fade out after the chorus.
    And so forth.
    Last song intro, verse, last chorus, and fade out applause.

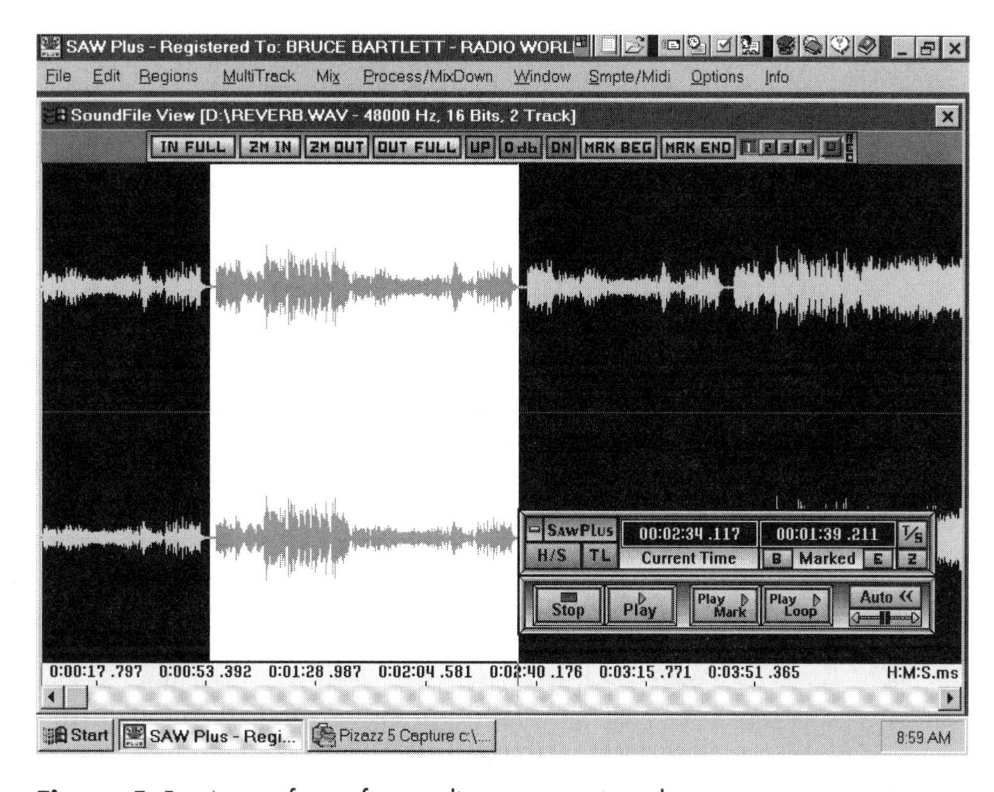

Figure 5–1 A waveform of an audio program viewed on a computer monitor. A region is highlighted.

MARKING THE REGIONS IN A GIG DEMO RECORDING

SONG 1 1ST PART
SONG 2 1ST PART
SONG 3 1ST PART
SONG 4 1ST PART
SONG 4 2ND PART
LEVEL
1 2 3 4-1 4-2
APPLAUSE
TIME →
APPLAUSE

LEVEL
FADE IN
1 2 3 4-1 4-2
FADE-OUT
APPLAUSE
TIME →
APPLAUSE

THE EDITED DEMO (THE PLAYLIST) WITH FADES ADDED

Figure 5–2 Top: Marking the regions in a gig demo recording. Bottom: The edited demo (the playlist) with fades added.

3. Now you will edit your demo to fit that format. Let's say you've already marked the beginning of the first song. Now place the cursor about 5 seconds after the first chorus, and mark that as your end point. The start and end points define a "region" or a section of the recording that you want to keep. Save the region you just marked as *Song 1* or whatever. Later, you will fade out the end of the song just after the chorus.
4. Put the cursor at the beginning of the next song you want to demo. Mark that as the start point. Then place the cursor about 5 seconds after the first chorus, and mark that as the end point. Save the region as Song 2. Mark and save other songs the same way.
5. You want the demo to end with wild applause. So, when you edit the last song, you will keep the ending of the song and about 15 seconds of applause afterwards. By *last song,* I mean the last song you want to play on your edited tape. This is not necessarily the last song played at the gig. Mark the last song like this. Place the cursor at the beginning of the song and mark the start point. Now place the cursor at the beginning of the first chorus and mark the end point. Zoom way in so you can mark this point with precision, just before a beat. Save the region as *Last Song Part 1* or whatever. Play through the song until you reach the beginning of the last chorus before the song ends. Place the cursor there and mark it as the start point. Finally, place the cursor 15 seconds into the applause after the song, and mark that as the end point. Save the region as *Last Song Part 2* or whatever.

Create a Playlist

The time has come to put all the songs end to end. In your playlist or on a stereo pair of tracks place song 1, followed by song 2, song 3, … , up through *Last Song Part 1* and *Last Song Part 2.* If you defined a region of crowd noise, put that first.

Then, play the transition from Part 1 to Part 2 in the last song. If your timing was correct, it should sound like a single, nonstop song. If not, you can go back and fine tune the edit points.

Add Fades and EQ

1. Now that the playlist is assembled, you can add the fades. Using your editor's fade function, fade into the beginning applause. Start from silence (pull the fader way down) and fade up over 4 seconds or so (Figure 5–2).

2. At the end of Song 1's chorus, you have about 5 seconds to fade out to silence.
3. Start Song 2 at full volume. Fade the end of Song 2, start Song 3 at full volume, and so on (Figure 5–2).
4. At the end of Last Song Part 2, let the applause run for about 5 seconds, then slowly fade it out over about 10 seconds. Those fade times are just suggestions—use your ears and do what sounds right. If your DAW has equalization, you might want to add some to the program.
5. Once the program is edited as you want it, copy it in real time onto a new DAT tape. This is your finished gig demo. Copy it onto cassettes.

Editing a Full-Length Gig Tape

Suppose you're editing a full-length gig tape in which you want to hear all of each song. The only editing you need is to remove songs that weren't played well and long pauses or mic feedback between songs.

In general, you will mark sections of stuff you want to keep. You will save these sections and put them end to end in a playlist.

Mark the first song about 5 seconds early so you can fade up crowd noise later. Now place the cursor at the beginning of stuff you want to throw out and mark that as the *end* point. Save the region you just marked as *Region 1* or whatever.

Mark just before the beginning of the next song you want to keep. Then place the cursor at the beginning of material you want to delete and mark that as the *end* point. Save the marked region as *Region 2* or whatever, and so on.

Put all the regions end to end in a playlist and check each transition point. If done well, the edited program should sound like a single continuous concert. Good luck!

6

TIPS ON BUILDING A RECORDING TRUCK

The ultimate way to record on location is with a dedicated recording van, bus, or truck. The truck contains your console, monitors, and recorders, so you don't need to cart them into the venue. It is a convenient way to work and saves setup time. Also, a truck provides a quiet, controlled monitoring environment.

You might take a remote truck to musicians' homes. That way the musicians don't have to disassemble their home MIDI setups.

When your truck is not on a job, you can pull it up to your studio and record or mix in the truck. Gear not being used for remotes can be used for studio work.

Depending on the kind of recording you want to do, you have several options of mobile recording services. Here are some options, ranked from simple to complex (cheap to expensive):

1. Carry MDMs into the venue to plug into the PA console. Let the PA system be your monitor system.
2. Carry MDMs, a mixer, and monitors into the venue. Set them up in an isolated room, taking your feeds from a splitter.
3. Install the MDMs and a mixer in a small van. Monitor with headphones. Run a snake and power cables to the stage.
4. Outfit a bus or semitrailer with recorders, mixer, and monitors.
5. Add signal processors so you can do complete mixes for clients or broadcasts.

6. Add a producer's lounge with a desk, phone, coffee maker, and microwave.
7. Add video monitors and satellite feeds.

Remote Truck Acoustics

The two main goals in designing your truck's acoustics are to attenuate the outside noises and to make the internal acoustics dead.

How to Attenuate the Outside Noises

Each door is a major sound leak, so seal the doors well with weather stripping.

Build up mass on the walls and floor. Attach two or three layers of particle board, intermixed with two layers of $\frac{1}{16}$-inch lead sheet or acoustical vinyl sheet. The total wall thickness might be 6 inches. The ceiling can be thinner. You might make the floor of two layers of $\frac{5}{8}$-inch tongue-and-groove wood glued together, floating on a layer of neoprene rubber. The truck suspension helps the isolation as well.

Here is a suggested treatment for a tour bus: Cover wall surfaces with $\frac{1}{2}$- or $\frac{3}{4}$-inch plywood secured only at the outer edges. This lets the panels vibrate and absorb low frequencies. Put fiberglass insulation behind the panels. Cover the windows with well-sealed $\frac{3}{4}$-inch plywood.

For more isolation, line the floor and engine wall with $\frac{1}{4}$-inch lead or acoustical vinyl sheeting. Seal the floor airtight.

How to Treat the Interior Acoustics

Make the room dead. Put as much absorption in the room as you can. On the walls, attach sound-absorbing material such as Sonex acoustic foam. Or obtain packing blankets, thick curtains, or 2-inch-thick absorbers. Hang them 3–4 inches from the walls. Spacing the absorber from the nearby surface extends absorption into the low frequencies.

It's important to absorb low frequencies as well as high frequencies. In the corners and on walls, mount bass traps such as ASC tube traps.

You might splay or angle the walls a little to avoid parallel surfaces.

To reduce air-conditioning noise, have the air exchange unit draw in air through the patchbay holes. This also vents heat from power amps placed low in the rack. Another solution is to place the air conditioner and fan-cooled amps in the driver's cab for isolation.

In a small room such as a truck interior, the sweet spot (where the monitors sound accurate) will be limited.

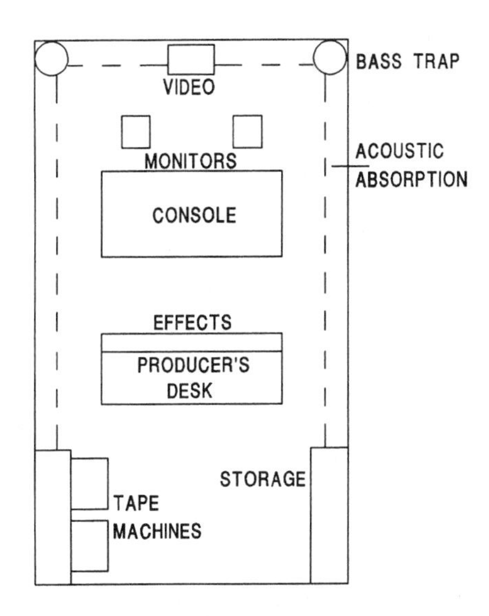

Figure 6–1
Example of a mobile recording truck layout.

Figure 6–1 shows the layout of a typical small recording truck.

Monitoring

Use Nearfield or midfield monitoring. Rather than putting the speakers on top of the console meter bridge, place them 1 or 2 feet behind the console on stands. This eliminates a console sound reflection, which causes peaks and dips in the frequency response. Moving the speakers back flattens the response.

If you equalize your monitors, try to measure them with a transfer-function analyzer, such as Meyer SIMM, MLSSA, JBL-Smaart Pro, or TEF. Use as little EQ as possible, and don't bother equalizing out the narrow-band notches.

Other Tips

- Arrange equipment symmetrically left and right, fore and aft, to distribute the weight evenly.
- Bolt down everything.
- On the road, connectors and cable harnesses are subject to damaging vibration. So screw some P-shaped clamps on walls and other points to support the cabling and remove weight from the connectors.
- Relieve the strain on the internal wiring in racks: Tie audio cables to one side and AC cables to the other side.

- Equipment bouncing during travel can loosen screws. Periodically tighten all the screws in your equipment, racks, and casters. Align analog tape machines when you arrive at your destination.
- Vacuum or blow the dust out of your gear, especially in power-amp air filters. Clean contacts and insert the plugs a few times to increase their reliability.
- For computers, consider getting an uninterruptible power supply (UPS), which provides electrical service in case of a power failure.
- Connect the snake to a multipin connector in an external bulkhead. Or, to avoid bad weather, bring the snake into the driver's cab and connect there.
- Finally, keep the truck height moderate, for example 11 feet 6 inches, so that it can fit almost anywhere.

The Record Plant Truck

To illustrate the construction of a high-end remote truck, let's describe a truck used by the Record Plant.

Construction

The truck weighs 65,000 pounds because it is so solid. Since the walls are high density, they don't flex as in normal truck bodies. The floor includes 3 inch steel I-beams. All ports are well sealed. The trim, made by a cabinetmaker, is solid cherry.

Areas in the Truck

- Producer's area.
- Mixer's suite.
- Tape machine room.
- Studio B, used for overdubs, live satellite transmissions, and live radio broadcasts.

Control Room Equipment

- Neve VRM console with Flying Faders. The one-button snapshot recall is handy for multiple-act live shows.
- KRK monitors.
- Video monitors and remote cameras.

- Power amps.
- Patch panel using $\frac{1}{4}$-inch TRS connectors. There are 140 mic lines with individual phantom power, and 3000 other patch points. The patch lines to the console are in groups of 24.

Machine Room Equipment

- Up to five Sony 48-track machines with Dolby SR.
- Patch bay including sync, telephone lines, satellite links, video, misc.
- D/A and A/D converters.
- Small video machines, cassette deck, tuner, DAT machines, Sony 1630 processor, and DMR 4000 time-code DATs.

Electrical System

- Electrical distribution in ceiling.
- Star grounding system.
- Breaker panel.
- Separate isolation transformers: 20 kVA on technical equipment, 10 kVA on each air conditioner and lights, 7.5 kVA for the backroom.

The Effanel Recording Truck

Owner-engineer Randy Ezratti describes the Effanel Music Mobile Unit L7:

> This is the world's first mobile recording studio whose control-room walls expand outward to create a spacious, studiolike environment. Its 14 foot width and 10 foot ceiling resemble some of the world's finest stationary control rooms.

The 48-foot expandable trailer has less isolation from outside noises than a standard truck because the seams and hinges let sound through. But when the truck is closed it has two walls, and so is quieter. Figure 6–2 shows the truck interior.

Equipment List

- Neve Capricorn 24-bit digital console, capable of 5.1 surround mixing. Console includes 80 mic preamps and 256 signal paths.
- Optional outboard mic preamps on stage: 52 channels of John Hardy M-1 preamps with Jensen transformers with all-discrete electronics; 24 channels of Millenia Media preamps.

Figure 6–2 Interior of Effanel mobile recording truck L7.

- Tape recorders (loaded per project):
  One Sony 3348 (upgraded) 48-track digital (16 bit or 24 bit) with DMU3348 external meter bridge.
  Two Otari MTR90/2 23-track analog with Dolby SR.
  Six TASCAM DA-88, 8-track digital with Neve 24-bit, 24-track option.
  One TASCAM DA-88, 8-track digital (for mixdown).
  Two Panasonic 3700 DATs.
  Two TASCAM cassettes decks.
- Monitor speakers: Meyer HD-1, Yamaha NS-10 with Bryston 4B power amplifier.
- Various outboard equipment (loaded per project).

Comments

Instant recall on the console is a very useful feature. You can show artists a quick mix after rehearsal, then go back to another console setup.

Effanel offers different technology levels for different projects, from soloists to full bands, from living rooms to stadiums.

Effanel's system philosophy is to keep it clean, keep it simple, keep the signal path as straight and short as possible.

Whenever possible, Effanel puts mic preamps on stage. A skilled technician sits on stage watching preamp levels. Preamps also can be remote controlled from the truck. The John Hardy mic preamps feed A/D converters. The converters feed a Telecast fiber-optic cable that runs from the stage to the truck. Because the fiber-optic signal is light waves, there are no problems with ground loops. The truck is completely isolated from the PA.

Randy feels that a conventional 250 foot snake degrades the mic signal compared to the fiber-optic link. (Some engineers, however, use a 250 foot snake feeding high-quality mic preamps at the console and get excellent results.) Using fiber-optics lets Effanel run a 1600 foot thin cable with no losses.

According to Randy, a recording made by taking inserts out of the PA console into MDMs results in a tape that is inadequate for commercial release. For major artists, he prefers to feed the tape machines directly from mic preamps on stage.

Doing live submixes in the truck can be difficult if outside noises interfere. So Randy tries to put each instrument or vocal on its own track.

Seeing how the top pros use a remote truck may inspire you to build your own. Whether you use a truck or just hand carry gear into a venue, I hope that you enjoy the thrill of on-location recording.

References

Much of the information in this chapter was derived from the workshop Tales from the Truck, which took place at the 103rd Convention of the Audio Engineering Society in New York (September 1997). The panelists were

Hamilton ("Ham") Brosius.

David Hewitt (of Record Plant) (on video).

Randy Ezratti (of Effanel).

Kooster McAllister (of Record Plant).

Steve Remote (of Aurasonic).

John Storyk (acoustic consultant).

Other valuable sources of information were

Dennis DeCamillo. "Room to Move," *EQ* (February 1998), pp. 70–78.

Gordon Jennings. "Maintaining Your Gear." *Gig* (April 1998), p. 118.

David Kelln. "On the Road Again." *Recording* (May 1998), pp. 18–30.

Part 2
Classical Music Recording and Stereo Microphone Techniques

7

MICROPHONE POLAR PATTERNS AND OTHER SPECIFICATIONS

Before we can understand how stereo microphone arrays work, we need to understand microphone polar patterns. These are explained in this chapter, as are other specifications that will help you choose appropriate microphones and accessories for stereo recording.

Polar Patterns

Microphones differ in the way they respond to sounds coming from different directions. Some respond the same to sounds from all directions; others have different output levels for sources at different angles around the microphone.

This varying sensitivity versus angle can be graphed as a polar pattern or polar response (see Figure 7–1). Polar patterns are plotted on polar graph paper as follows: In an anechoic chamber, the microphone is exposed to a tone of a single frequency and its output voltage is measured as it is rotated around its diaphragm. The voltage at 0° (on-axis) is called the *0 dB reference*, and the voltages at other angles are referenced to that. In other words, the polar-response graph plots relative sensitivity in dB versus angle of sound incidence in degrees. Often, several such plots are made at various frequencies. Note that the microphone rotation should be clockwise if the degree marks increase in a counterclockwise direction.

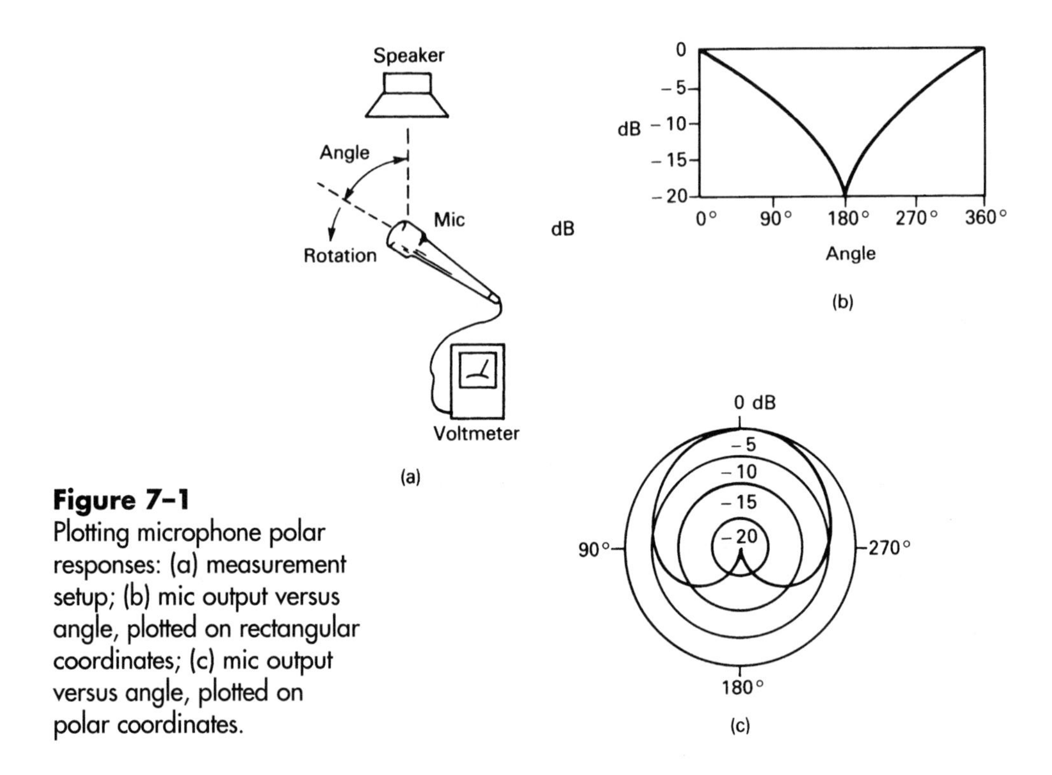

Figure 7–1
Plotting microphone polar responses: (a) measurement setup; (b) mic output versus angle, plotted on rectangular coordinates; (c) mic output versus angle, plotted on polar coordinates.

Another way to generate a polar plot is by using time delay spectrometry. Measure the microphone's frequency response every 10° around the microphone, then process the data with a program that draws a polar plot at selected frequencies.

The three major polar patterns are omnidirectional, unidirectional, and bidirectional. An omnidirectional microphone is equally sensitive to sounds arriving from all directions. A unidirectional microphone is most sensitive to sounds arriving from one direction (in front of the microphone) and rejects sounds entering the sides or rear of the microphone. A bidirectional microphone is most sensitive to sounds arriving from two directions (in front of and behind the microphone) but rejects sounds entering the sides.

Three types of unidirectional patterns are cardioid, supercardioid, and hypercardioid. A microphone with a cardioid pattern is sensitive to sounds arriving from a broad angle in front of the microphone. It is about 6 dB less sensitive at the sides and about 15 to 25 dB less sensitive at the rear. To hear how a cardioid pickup pattern works, talk into a cardioid microphone from all sides while listening to its output. Your reproduced voice will be loudest when you talk into the front of the microphone and softest when you talk into the rear.

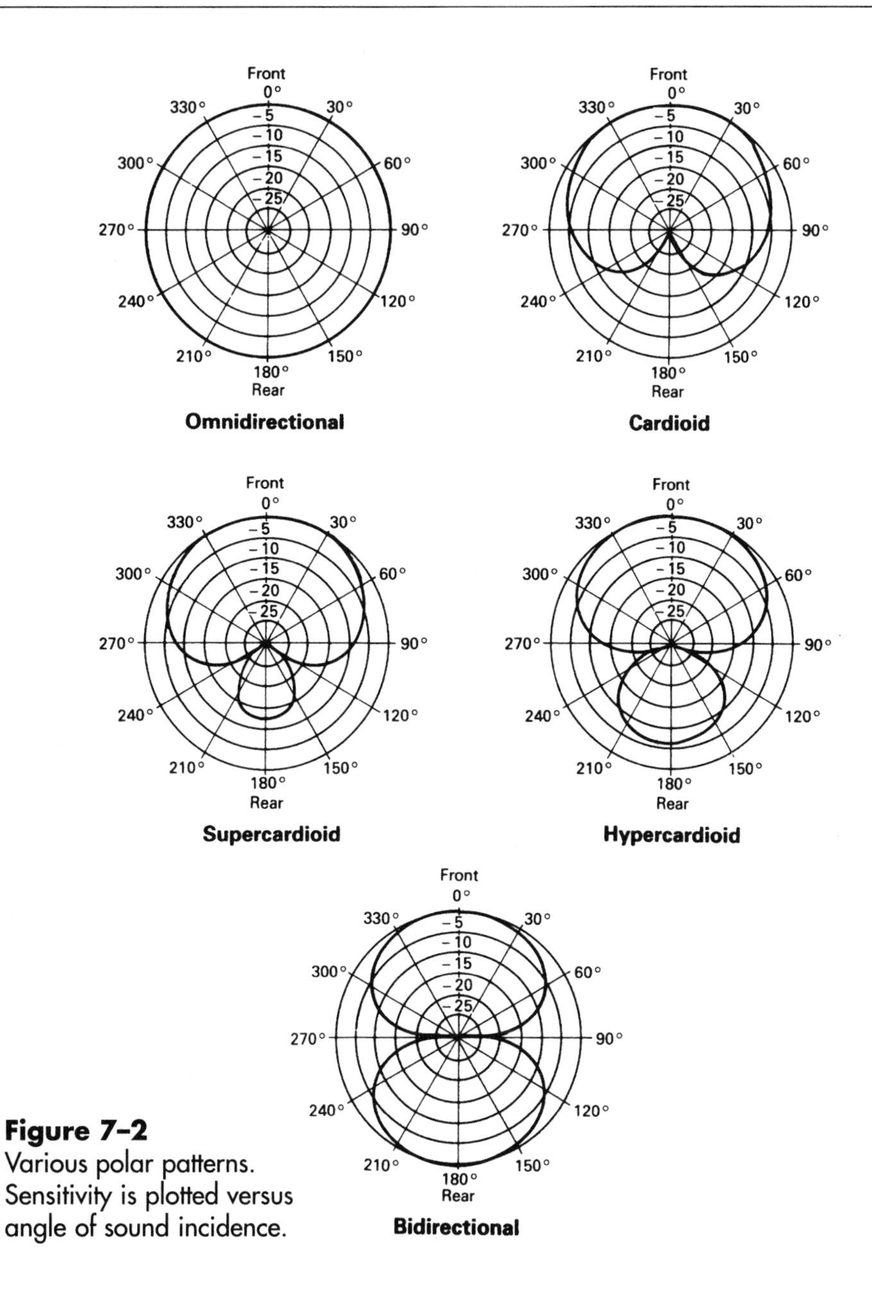

Figure 7–2
Various polar patterns. Sensitivity is plotted versus angle of sound incidence.

The supercardioid pattern is 8.7 dB down at the sides and has two nulls (points of least pickup) at 125° off axis. *Off axis* means "away from the front." The hypercardioid pattern is 12 dB down at the sides and has two nulls of least pickup at 110° either side off axis.

Figure 7–2 shows various polar patterns. Note that a polar plot is not a geographical map of the "reach" of a microphone. A microphone does not suddenly become dead outside its polar pattern; there is no "outside." The graph merely plots sensitivity at one frequency as distance from the origin; it is not the spatial spread of the pattern.

Advantages of Each Pattern

Omnidirectional microphones have several characteristics that make them especially useful for certain applications. Use omnidirectional microphones when you need

- All-around pickup.
- Extra pickup of room reverberation.
- Low pickup of mechanical vibration and wind noise.
- Extended low-frequency response (in condenser mics).
- Lower cost in general.
- Freedom from proximity effect (up-close bass boost).

Use directional microphones when you need

- Rejection of room acoustics and background noise.
- Coincident or near-coincident stereo (explained in the next chapter).

Other Polar Pattern Considerations

In most microphones, it is desirable that the polar pattern stay reasonably consistent at all frequencies. If not, you'll hear off-axis coloration: The mic will sound tonally different on and off axis. Uniform polar patterns at different frequencies indicate similar frequency responses at different angles of incidence.

Some condenser mics come with switchable polar patterns.

An omnidirectional boundary microphone (a surface-mounted microphone, explained later) has a half-omni, or hemispherical, polar pattern. A unidirectional boundary microphone has a half-supercardioid or half-cardioid polar pattern. The boundary mounting increases the directionality of the microphone, thus reducing pickup of room acoustics.

Transducer Type

We've seen that microphones differ in their polar patterns. They also differ in the way they convert sound to electricity. The three operating principles of recording microphones are condenser, moving coil, and ribbon.

In a condenser microphone (shown in Figure 7–3), a conductive diaphragm and an adjacent metallic disk (called a *backplate*) are charged

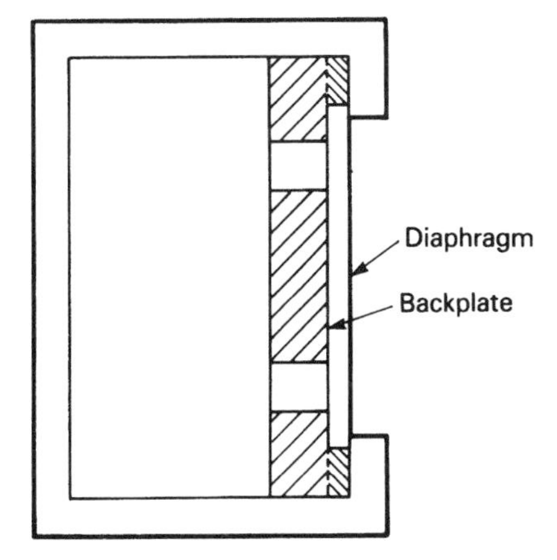

Figure 7–3
Condenser microphone.

with static electricity to form a capacitor. Sound waves vibrate the diaphragm; this vibration varies the capacitance and produces a voltage similar to the incoming sound wave. The diaphragm and backplate can be charged (biased) by an external power supply, or they can be permanently charged by an electret material in the diaphragm or on the backplate. *Capacitor* is the modern term for "condenser," but the name *condenser microphone* has stuck and is popular usage. Some condenser mics use variable capacitance to frequency modulate a radio-frequency carrier.

The condenser microphone is the preferred type for stereo recording because it generally has a wide, smooth frequency response; a detailed sound quality; and high sensitivity. It requires a power supply to operate, such as a battery or phantom power. Phantom power is provided by an external supply or from the recording mixer.

In a moving-coil microphone (sometimes called *dynamic*), a voice coil attached to a diaphragm is suspended in a magnetic field (as shown in Figure 7–4). Sound waves vibrate the diaphragm and voice coil, producing an electric signal similar to the incoming sound wave. No power supply is needed.

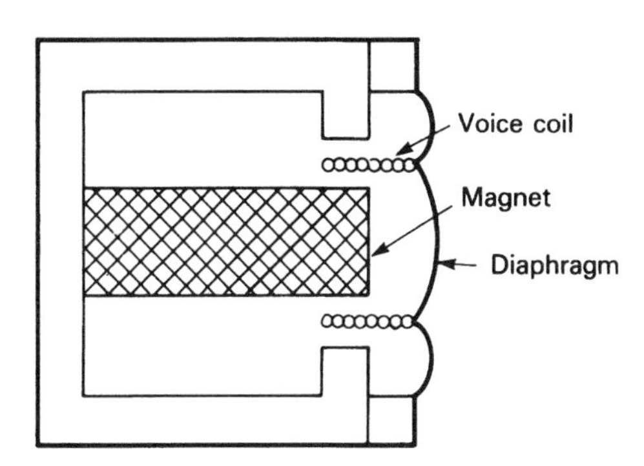

Figure 7–4
Moving-coil microphone.

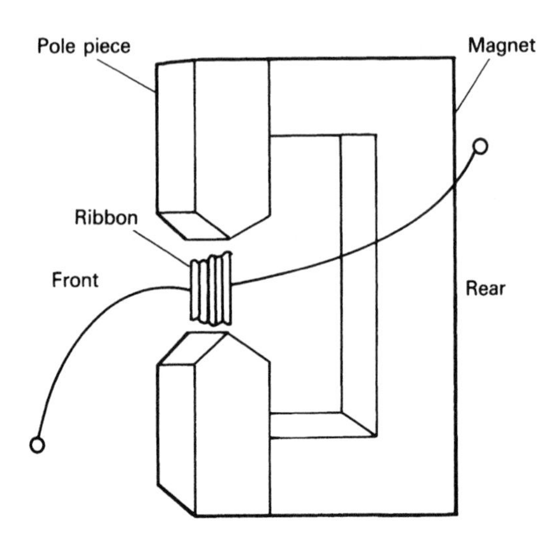

Figure 7–5
Ribbon microphone.

In a ribbon microphone (Figure 7–5), the conductor is a thin metallic strip, or ribbon. Sound waves vibrate the ribbon, which generates electricity. A printed ribbon microphone has a thin plastic diaphragm with an implanted ribbon.

Since the moving-coil microphone generally has a rougher frequency response and lower sensitivity than the condenser type, the moving-coil type is used less often for stereo recording. The ribbon, however, has a smooth response and has been used in Blumlein arrays (crossed bidirectionals).

Maximum Sound Pressure Level

Another microphone specification is maximum sound pressure level (SPL). To clarify this specification, first we need to explain the term. SPL is a measure of the intensity of a sound. The quietest sound we can hear, the threshold of hearing, measures 0 dB SPL. Normal conversation at a distance of one foot measures about 70 dB SPL; painfully loud sound is above 120 dB SPL.

Maximum SPL is the sound pressure level at which a microphone's output signal starts to distort; this is usually the SPL at which the microphone produces 3 percent total harmonic distortion (THD). (Some manufacturers use 1 percent THD.) If a microphone has a maximum SPL spec of 125 dB SPL, that means the microphone starts to distort audibly when the sound pressure level produced by the source reaches 125 dB SPL. A maximum SPL spec of 120 dB is good, 135 dB is very good, and 150 dB is excellent.

Sensitivity

Sensitivity is another specification to consider. It is a measure of the efficiency of a microphone. A very sensitive microphone produces a relatively high output voltage for a sound source of a given loudness.

Microphone sensitivity is often stated in millivolts/Pa, where 1 Pa = 1 pascal = 94 dB SPL. This spec tells what voltage the microphone produces (in millivolts) when picking up a 1000 Hz tone at 94 dB SPL. An older sensitivity spec is "dB re 1 volt per microbar (dBV/μbar)."

The following list gives typical sensitivity specs for the three microphone types.

Condenser: 6 mV/Pa (high sensitivity)

Moving coil: 2 mV/Pa (medium sensitivity)

Ribbon or small moving coil: 1 mV/Pa (low sensitivity)

Differences of a few dB among microphones are not critical.

The sensitivity of a microphone doesn't affect its sound quality. Rather, sensitivity affects the audibility of mixing console noise (hiss). To achieve the same recording level, a low-sensitivity mic requires more mixer gain than a high-sensitivity mic, and more gain usually results in more noise. If you record quiet, distant instruments such as a classical guitar or chamber music, you'll hear more mixer noise with a low-sensitivity mic than with a high-sensitivity mic, all else being equal. Since stereo miking usually is done at a distance, high sensitivity is an asset.

Sensitivity sometimes is called *output level,* but the two terms are not synonymous. Sensitivity is the output level produced by a particular input sound pressure level. The higher the SPL, the higher is the output of any microphone.

Self-Noise

Self-noise is the electrical noise (hiss) a microphone produces. The microphone is put in a soundproof box, and its output noise voltage is measured. Self-noise is specified as the dB SPL of a source that would produce the same output voltage as the noise.

The self-noise spec usually is A weighted; that is, the noise was measured through a filter that makes the measurement correlate more closely with the annoyance value. The filter rolls off low and high frequencies to simulate the frequency response of the human ear. An A-weighted self-noise spec of 14 dB SPL or less is excellent (quiet); a spec around 20 dB SPL is very good; and a spec around 30 dB SPL is fair.

Signal-to-Noise (S/N) Ratio

Although referred to as a *ratio,* this is the difference between SPL and self-noise, all in dB. The higher the SPL of the sound source at the mic or the lower the self-noise of the mic, the higher is the S/N "ratio." For example, if the sound pressure level is 94 dB at the microphone and the mic self-noise is 24 dB, the S/N ratio is 70 dB.

The higher the S/N ratio, the cleaner (more noise free) is the signal. Given an SPL of 94 dB, a signal-to-noise ratio of 74 dB is excellent; 64 dB is good.

Microphone Types

Another specification is the type of microphone, the generic classification. Some types of microphones for stereo miking are free field, boundary, stereo, shotgun, and parabolic.

Free-Field Microphone

Most microphones are of this type. They are meant to be used in a free field; that is, away from reflective surfaces.

Boundary Microphone

A boundary microphone is designed to be used on a surface such as a floor, wall, table, piano lid, baffle, or panel. One example of a boundary microphone is the Crown Pressure Zone Microphone (PZM) (shown in Figure 7–6). It includes a miniature electret condenser capsule mounted face down next to a sound-reflecting plate or boundary. Because of this construction, the microphone diaphragm receives direct and reflected sounds in phase at all frequencies, avoiding phase interference between them. The claimed benefits are a wide, smooth frequency response free of phase cancellations, excellent clarity and "reach," a hemispherical polar pattern, and uniform frequency response anywhere around the microphone. Because of this last characteristic, hall reverberation is picked up without tonal coloration.

If an omnidirectional boundary mic is placed on a panel, it becomes directional. Thus, boundary mics on angled panels can be used for stereo arrays. Boundary microphones also are available with a unidirectional polar pattern. They have the benefits of both boundary mounting and the unidirectional pattern. Such microphones are well suited for stage-floor pickup of drama, musicals, or small musical ensembles.

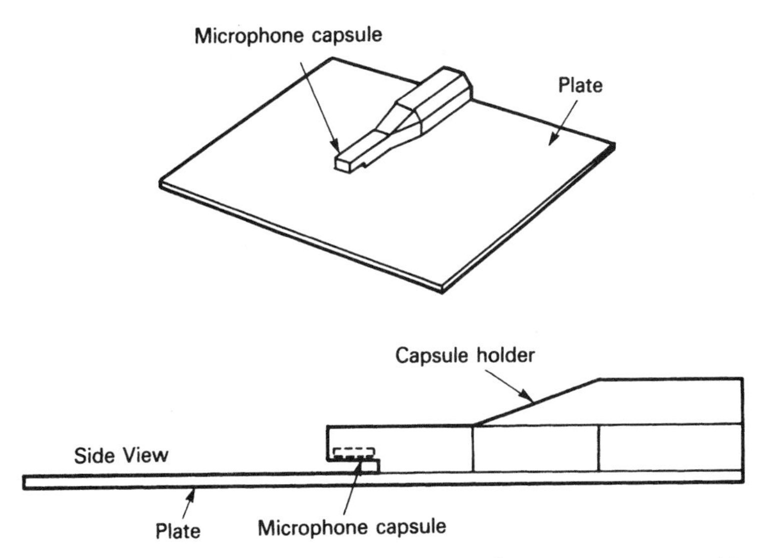

Figure 7–6 Crown PZM construction. (Courtesy of Crown International.)

Stereo Microphone

A stereo microphone combines two mic capsules in a single housing for convenient stereo recording. Simply place the microphone about 10–15 feet in front of a band, choir, or orchestra, and you'll get a stereo recording with little fuss. Several models of stereo microphones are listed in Chapter 16; one is shown in Figure 7–7.

Most stereo microphones are made with coincident microphone capsules. The absence of horizontal spacing between the capsules means the absence of any delay or phase shift between their signals. Thus, the microphone is mono-compatible: The frequency response is the same in mono as in stereo, because there are no phase cancellations if the two channels are combined.

Stereo microphones are available in many configurations, such as XY, MS, Blumlein, ORTF, OSS, Soundfield, and SASS (these types are described

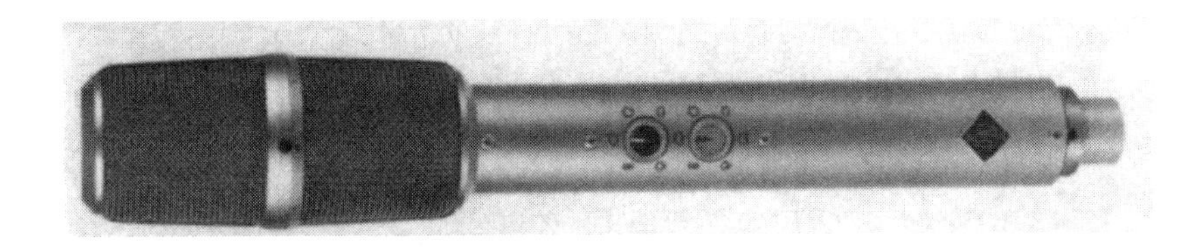

Figure 7–7 Stereo microphone, a Neumann USM 69. (Courtesy of Gotham Audio Corporation.)

in Chapters 10 and 11). Unlike the other types, the MS stereo microphone and Soundfield microphone let you remote-control the stereo spread and vary the stereo spread after recording. In general, a stereo microphone is easier to set up than two separate microphones but more expensive.

Shotgun Microphone

A shotgun or line microphone is a long, tube-shaped microphone with a highly directional pickup pattern. It is used for highly selective pickup or maximum rejection of background noise. Typical uses are for film and video dialogue, news gathering, and outdoor recording. A stereo shotgun microphone is a mid-side type with a shotgun for the middle element and bidirectional for the side element.

Parabolic Microphone

This type of microphone has a large parabolic dish, or reflector, to focus sound on the microphone element. The parabolic type is even more directional than the shotgun type but has a rougher, narrower frequency response. Microphones with big parabolic reflectors are directional down to lower frequencies than microphones with small reflectors. Also, the bigger the reflector, the lower the frequency response extends. No stereo parabolic microphones are commercially available, but environmental recordist Dan Gibson has constructed his own to make some excellent recordings.

Microphone Accessories

Various accessories used with microphones enhance their convenience, aid in placement, or reduce vibration pickup.

Stands and Booms

These adjustable devices hold the microphones and let you position them as desired. A microphone stand has a heavy metal base that supports a vertical pipe. At the top of the pipe is a rotating clutch that lets you adjust the height of a smaller telescoping pipe inside the larger one. The top of the small pipe has a standard $\frac{5}{8}$-inch-27 thread, which screws into a microphone stand adapter. Camera stores have photographic stands, which are collapsible and lightweight—ideal for recording on-location. The thread usually is $\frac{1}{4}$-inch-20, which requires an adapter to fit a $\frac{5}{8}$-inch-27 thread in a mic stand adapter.

A boom is a long pipe that attaches to a mic stand. The angle and length of the boom are adjustable. The end of the boom is threaded to accept a microphone stand adapter, and the opposite end is weighted to balance the weight of the microphone. You can use a boom to raise a microphone farther off the floor in order to stereo-mike an orchestra.

Stereo Microphone Adapter

A stereo microphone adapter, stereo bar, or stereo rail mounts two microphones on a single stand for convenient coincident and near-coincident stereo miking. Several models of these are listed in Chapter 16, and one is shown in Figure 7–8. In most models, the microphone spacing and angling are adjustable.

Shock Mount

This device mounts on a microphone stand and holds a microphone in a resilient suspension to isolate it from mechanical vibrations such as stand and floor thumps. The shock mount acts as a spring that resonates at a subaudible frequency with the mass of the microphone. This mass-spring system attenuates mechanical vibrations above its resonance frequency.

Many microphones have an internal shock mount that isolates the microphone capsule from its housing; this reduces handling noise as well as stand thumps.

Figure 7–8 Stereo microphone adapter, a Schoeps UMS20. (Courtesy of Posthorn Recordings.)

Phantom Power Supply

Condenser microphones need power to operate their internal circuitry. Some use a battery; others use a remote phantom power supply. This supply can be a stand-alone box, which you connect between the mic and mixer mic input. Some mixing consoles have phantom power built in, available at each mic connector. Phantom power usually is 48 V DC on pins 2 and 3, with respect to pin 1.

Phantom power is supplied to the mic through its two-conductor shielded audio cable. The microphone receives power from and sends audio to the mixer along the same cable conductors.

Junction Box and Snake

If you're recording with more than three microphones, you might want to plug them into a junction box with multiple connectors. A single, thick multiconductor cable, called a *snake,* carries the signals from the junction box to your mixer. At the mixer end, the cable divides into several mic connectors that plug into the mixer.

Splitter

You might need to send your microphone signals simultaneously to a broadcast mixer, recording mixer, and sound-reinforcement mixer. A microphone splitter does the job. It has one input for each microphone and two or three isolated outputs per microphone to feed each mixer. This device is passive; a distribution amplifier with the same function is active (it has amplification).

With a good grasp of microphone characteristics and accessories, we're now ready to discuss stereo miking theory.

Reference

This chapter is based on B. Bartlett, "Microphones," in *Practical Recording Techniques, Second Edition* (Boston: Focal Press, 1998), Chapter 6.

8

OVERVIEW OF STEREO MICROPHONE TECHNIQUES

Stereo microphone techniques are used mainly to record classical music ensembles and soloists on location. These methods capture a sonic event as a whole, typically using only two or three microphones. During playback of a stereo recording, images of the musical instruments are heard in various locations between the stereo speakers. These images are in the same places, left to right, that the instruments were at the recording session. In addition, true-stereo miking conveys

- The depth or distance of each instrument.
- The distance of the ensemble from the listener (the perspective).
- The spatial sense of the acoustic environment, the ambience or hall reverberation.

Why Record in Stereo?

When planning a recording session, you may ask yourself, "Should I record in stereo with just a few mics? Or should I use several microphones placed close to the instruments and mix them with a mixer?"

Stereo miking is preferred for classical music, such as a symphony performed in a concert hall or a string quartet piece played in a recital hall. For classical music recording, stereo methods have several advantages over close-mic methods.

For example, I said that stereo miking preserves depth, perspective, and hall ambience—all part of the sound of classical music as heard by the audience. These characteristics are lost with multiple close-up pan-potted microphones. But with a good stereo recording, you get a sense of an ensemble of musicians playing together in a shared space. Also, a pair of mics at a distance relays instrument timbres more accurately than close-up mics. Close-miked instruments in a classical setting sound too bright, edgy, or detailed compared to how they sound in the audience area.

Another advantage of stereo miking is that it tends to preserve the ensemble balance as intended by the composer. The composer has assigned dynamics (loudness notations) to the various instruments to produce a pleasing ensemble balance in the audience area. So, the correct balance or mix of the ensemble occurs at a distance, where all the instruments blend together acoustically. But this balance can be upset with multiple miking; you must rely on your own judgment (and the conductor's) regarding mixer settings to produce the composer's intended balance. Of course, even a stereo pair of mics can yield a faulty balance. But a stereo pair, being at a distance, is more likely to reproduce the balance as the audience hears it.

Other Applications for Stereo Miking

In contrast to a classical music recording, a pop music recording is made with multiple close mics because it sounds tighter and cleaner, which is the preferred style of production for pop music. Close miking also lets you experiment with multitrack mixes after the session. Still, stereo miking can be used in pop music sessions for large sound sources within the ensemble, such as groups of singers; piano; drum-set cymbals overhead; vibraphone, xylophone, and other percussion instruments; or string and horn sections.

Other uses for stereo miking are samples; sound effects; background ambience and stereo dialog for film, video, and electronic news gathering; audience reaction; sports broadcasts; radio group discussions; radio plays and other drama.

Goals of Stereo Miking

Let's focus now on stereo miking a large musical ensemble and determine what we want to achieve. One objective is accurate localization. That is, the reproduced instruments should appear in the same relative locations as they were in the live performance. When this is achieved, instruments in the center of the ensemble are accurately reproduced midway between the two playback speakers. Instruments at the sides of the ensemble are reproduced from the left or right speaker. Instruments located halfway to one side are reproduced halfway to one side, and so on.

Figure 8–1 shows three stereo localization effects. In Figure 8–1(a), various instrument positions in an orchestra are shown: left, left-center, center, right-center, right. In Figure 8–1(b), the reproduced images of these instruments are accurately localized between the stereo pair of speakers. The stereo spread or stage width extends from speaker to speaker. If the microphones are placed improperly, the effect is either the narrow stage width shown in Figure 8–1(c) or the exaggerated separation shown in Figure 8–1(d). (Note that a large ensemble should spread from speaker to speaker, while a quartet can have a narrower spread.)

To judge these stereo localization effects, it's important to position yourself properly with respect to the monitor speakers. Sit as far from the speakers as they are spaced apart. The speakers will appear to be 60° apart, which is about the same angle an orchestra fills when viewed from the typical ideal seat in the audience (say, 10th row center). Sit exactly between the speakers (equidistant from them); otherwise, the images will shift toward the side on which you're sitting and become less sharp. Also, pull out the speakers several feet from the walls; this delays and weakens early reflections, which can degrade stereo imaging.

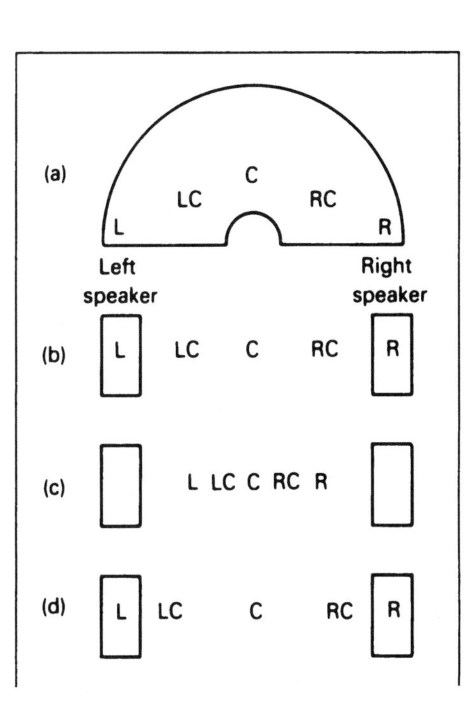

Figure 8–1
Stereo localization effects:
(a) orchestra instrument locations (top view);
(b) images accurately localized between speakers (the listener's perception);
(c) narrow stage-width effect;
(d) exaggerated separation effect.

The reproduced size of an instrument or instrumental section should match its size in real life. A guitar should be a point source; a piano or string section should have some stereo spread. Each instrument's location should be as clearly defined as it was in the concert hall, as heard from the ideal seat. Some argue that the reproduced images should be sharper than in real life to supplant the missing visual cues; in other words, since you can't see the instruments during loudspeaker reproduction, extra-sharp images might enhance the realism.

The reproduced reverberation (concert hall ambience) should either surround the listener, or at least it should spread evenly between the speakers (as shown in Figure 8–2). Typical stereo miking techniques reproduce the hall reverberation up front, in a line between the speakers, so you get no sense of being immersed in the hall ambience. To make the recorded reverberation surround the listener, you need extra speakers to the side or rear, an add-on reverberation simulator, or a head-related crosstalk canceler (explained in Chapter 12). However, spaced-microphone recordings artificially produce a sense of some surrounding reverberation.

There should also be a sense of stage depth. Front-row instruments should sound closer than back-row instruments.

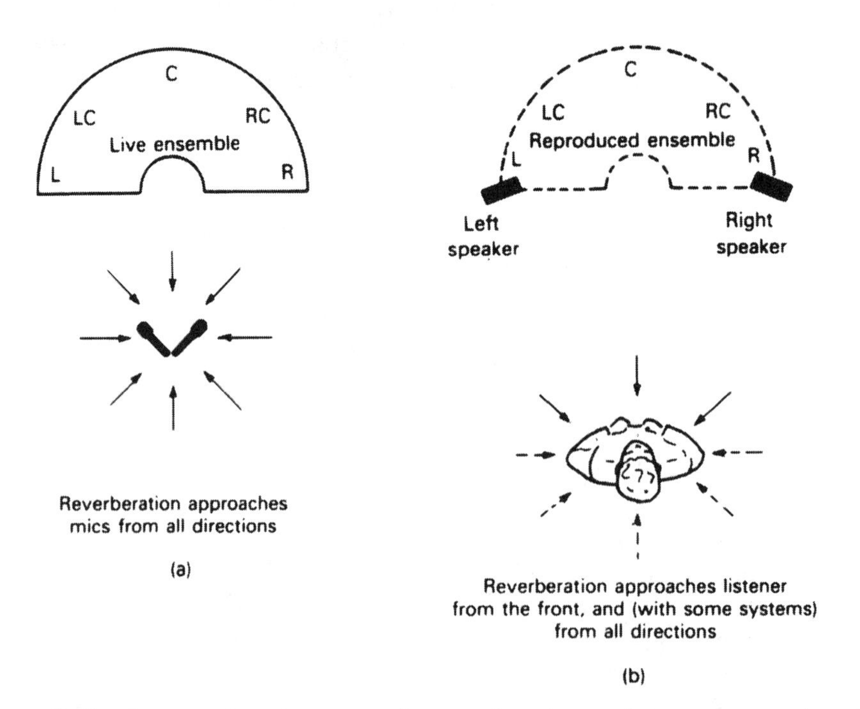

Figure 8–2 Accurate imaging—sound source location and size are reproduced during playback, as well as the reverberant field: (a) recording, (b) playback. (Courtesy of Focal Press.)

Types of Stereo Microphone Techniques

Three general microphone techniques commonly are used for stereo recording: the coincident pair, the spaced pair, and the near-coincident pair (Bartlett, 1979; Eargle, 1981; Streicher and Dooley, 1984). A fourth technique uses baffled omnis or an artificial head, covered in Chapter 12. Let's look at the first three techniques in detail.

Coincident Pair

With the coincident-pair method (the XY or intensity stereo method), two directional microphones are mounted with their grilles nearly touching and their diaphragms placed one above the other, angled apart to aim approximately toward the left and right sides of the ensemble (as in Figure 8–3). For example, two cardioid microphones can be mounted angled apart, their grilles one above the other. Other directional patterns can be used, too. The greater the angle between microphones and the narrower the polar pattern, the wider is the stereo spread.

Let's explain how the coincident-pair technique produces localizable images. As described in Chapter 7, a directional microphone is most sensitive to sounds in front of it (on axis) and progressively less sensitive to sounds arriving off axis. That is, a directional mic produces a relatively high-level signal from the sound source it's aimed at and a relatively low-level signal for all other sound sources.

The coincident-pair method uses two directional mics symmetrically angled from the center line, as in Figure 8–3. Instruments in the center of

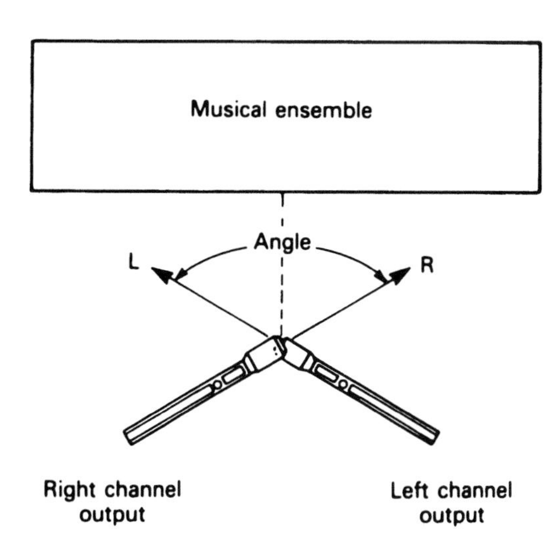

Figure 8–3
Coincident-pair technique.

the ensemble produce an identical signal from each microphone. During playback, an image of the center instruments is heard midway between the stereo pair of loudspeakers. That is because identical signals in each channel produce a centrally located image.

If an instrument is off center to the right, it is more on axis to the right-aiming mic than to the left-aiming mic; therefore, the right mic will produce a higher-level signal than the left mic. During playback of this recording, the right speaker will play at a higher level than the left speaker, reproducing the image off center to the right, where the instrument was during recording.

The coincident array codes instrument positions into level differences (intensity or amplitude differences) between channels. During playback, the brain decodes these level differences back into corresponding image locations. A pan pot in a mixing console works on the same principle.

If one channel is 15 to 20 dB louder than the other, the image shifts all the way to the louder speaker. So, if we want the right side of the orchestra to be reproduced at the right speaker, the right side of the orchestra must produce a signal level 20 dB higher from the right mic than from the left mic. This occurs when the mics are angled apart sufficiently. The correct angle depends on the polar pattern. Instruments partway off center produce interchannel level differences less than 20 dB, so they are reproduced partway off center.

Listening tests have shown that coincident cardioid microphones tend to reproduce the musical ensemble with a narrow stereo spread. That is, the reproduced ensemble does not spread all the way between speakers.

The Blumlein array is a coincident pair with excellent localization. It has two bidirectional mics angled 90° apart, facing the left and right sides of the ensemble.

A special form of the coincident-pair technique is the mid-side (MS) recording method illustrated in Figure 8–4. A microphone facing the middle of the orchestra is summed and differenced with a bidirectional microphone aiming to the sides. This produces left- and right-channel signals. With this technique, the stereo spread can be remote controlled by varying the ratio of the mid signal to the side signal. This remote control is useful at live concerts, where you can't physically adjust the microphones during the concert. MS localization accuracy is excellent.

Mid-side recordings are sometimes said to lack spaciousness. But this can be improved with spatial equalization, in which a circuit boosts the bass 4 dB (+2 dB at 600 Hz) in the L – R, or side, signal, with a corresponding cut in the L + R, or mid, signal (Griesinger, 1986). Another way to improve the spaciousness is to mix in a distant MS microphone, one set about 30–75 feet from the main MS microphone.

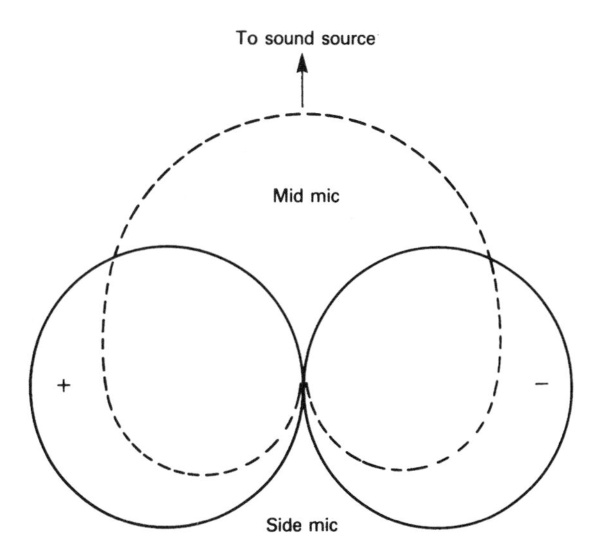

Figure 8–4
Mid-side technique. Left channel = mid + side. Right channel = mid − side. The polarity of the side mic lobes is indicated by plus and minus.

Two coincident microphone capsules can be mounted in a single housing for convenience: This forms a stereo microphone. Several of these are listed in Chapter 16.

A recording made with coincident techniques is mono-compatible; that is, the frequency response is the same in mono or stereo. Because of the coincident placement, there is no time or phase difference between channels to degrade the frequency response if both channels are combined to mono. If you expect your recordings to be heard in mono (say, on radio, TV, or film), you should consider coincident methods.

Spaced Pair

With the spaced-pair (or A–B) technique, two identical microphones are placed several feet apart, aiming straight ahead toward the musical ensemble (as in Figure 8–5). The mics can have any polar pattern, but the

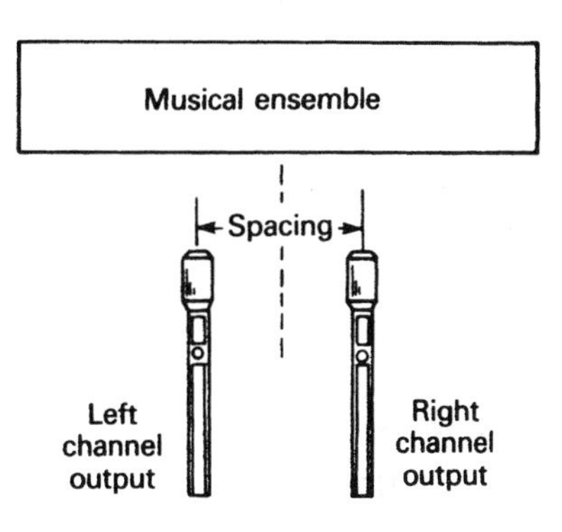

Figure 8–5
Spaced-pair technique.

omnidirectional pattern is the most popular for this method. The greater the spacing between the microphones, the greater is the stereo spread.

Instruments in the center of the ensemble produce an identical signal from each microphone. During playback of this recording, an image of the center instruments is heard midway between the stereo pair of loudspeakers.

If an instrument is off center, it is closer to one mic than the other, so its sound reaches the closer microphone before it reaches the other one. Consequently, the microphones produce an approximately identical signal, except that one mic signal is delayed with respect to the other. If you send an identical signal to two stereo speakers with one channel delayed, the sound image shifts off center. With a spaced-pair recording, off-center instruments produce a delay in one mic channel, so they are reproduced off center.

The spaced-pair array codes instrument positions into time differences between channels. During playback, the brain decodes these time differences back into corresponding image locations. It takes only about 1.5 milliseconds of delay to shift an image all the way to one speaker, so if we want the right side of the orchestra to be reproduced at the right speaker, its sound must arrive at the right mic about 1.5 milliseconds before it reaches the left mic. In other words, the mics should be spaced about 2 feet apart, because this spacing produces the appropriate delay to place right-side instruments at the right speaker. Instruments partway off center produce interchannel delays less than 1.5 milliseconds, so they are reproduced partway off center.

If the spacing between mics is, say, 12 feet, instruments slightly off center produce interchannel delays greater than 1.5 milliseconds, which places their images at the left or right speaker. This is called an *exaggerated separation* or *ping-pong effect.*

On the other hand, if the mics are too close together, the delays produced will be inadequate to provide much stereo spread. In addition the mics will tend to favor the center of the ensemble because the mics are closest to the center instruments.

To record a good musical balance, we need to place the mics about 10 or 12 feet apart, but such a spacing results in exaggerated separation. One solution is to place a third microphone midway between the original pair and mix its output to both channels. That way, the ensemble is recorded with a good balance and the stereo spread is not exaggerated.

The spaced-pair method tends to make off-center images relatively unfocused or hard to localize, for this reason: Spaced-microphone recordings have time differences between channels and stereo images produced solely by time differences are relatively unfocused. Centered instruments still are heard clearly in the center, but off-center instruments are difficult to pinpoint between speakers. This method is useful if you prefer the sonic images to be diffuse, rather than sharply focused (say, for a blended effect).

Spaced microphones pose another problem. The large time differences between channels correspond to gross phase differences between channels. Out-of-phase, low-frequency signals can cause excessive vertical modulation of a record groove, making records difficult to cut unless the cutting level or low-frequency stereo separation is reduced. (This is no problem with CDs or cassettes.) In addition, combining both mics to mono sometimes causes phase cancellations of various frequencies, which may or may not be audible.

There is an advantage with spaced miking, however. Spaced microphones are said to provide a warm sense of ambience, in which concert hall reverberation seems to surround the instruments and, sometimes, the listener. Here is why: The two channels of recorded reverberant sound are incoherent; that is, they have random phase relationships. Incoherent signals from stereo loudspeakers sound diffuse and spacious. Since reverberation is picked up and reproduced incoherently by spaced microphones, it sounds diffuse and spacious. The simulated spaciousness caused by the phasiness is not necessarily realistic (Lipschitz, 1986), but it is pleasant to many listeners.

Another advantage of the spaced-microphone technique is the ability to use omnidirectional microphones. An omnidirectional condenser microphone has more extended low-frequency response than a unidirectional condenser microphone and tends to have a smoother response and less off-axis coloration. (Microphone characteristics are explained in detail in Chapter 7.)

Near-Coincident Pair

As shown in Figure 8–6, the near-coincident technique uses two directional microphones angled apart, with their grilles horizontally spaced a few inches apart. Even a few inches of spacing increases the stereo spread and adds a sense of depth and airiness to the recording. The greater the angle or spacing between mics, the greater is the stereo spread.

Here is how this method works: Angling directional mics produces level differences between channels; spacing mics produces time differences. The interchannel level differences and time differences combine to create the stereo effect. If the angling or spacing is too great, the result is exaggerated separation. If the angling or spacing is too small, the result is a narrow stereo spread.

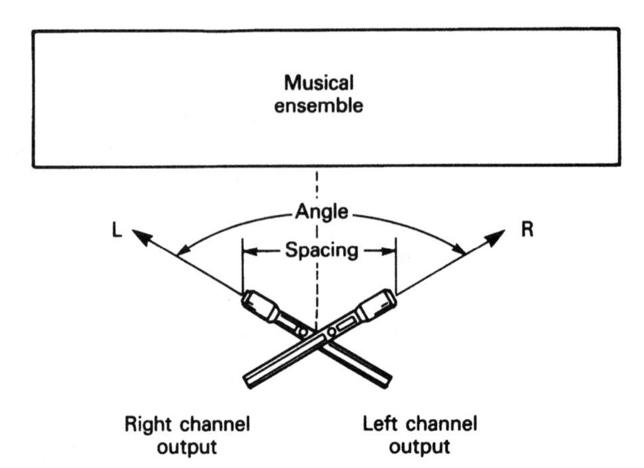

Figure 8–6
Near-coincident-pair technique.

The most common example of the near-coincident method is the ORTF system, which uses two cardioids angled 110° apart and spaced 7 inches (17 cm) apart horizontally (ORTF stands for Office de Radiodiffusion Television Française, French Broadcasting Organization). This method tends to provide accurate localization; that is, instruments at the sides of the orchestra are reproduced at or very near the speakers, and instruments halfway to one side tend to be reproduced halfway to one side.

Comparing the Three Stereo Miking Techniques

The coincident-pair technique has the following features:

- It uses two directional mics angled apart with grilles nearly touching, one mic's diaphragm above the other.
- Level differences between channels produce the stereo effect.
- Images are sharp.
- Stereo spread ranges from narrow to accurate.
- Signals are mono-compatible.

The spaced-pair technique has these features:

- It uses two mics spaced several feet apart.
- Time differences between channels produce the stereo effect.
- Off-center images are diffuse.
- Stereo spread tends to be exaggerated unless a third, center mic is used.
- It provides a warm sense of ambience.
- It may cause record-cutting problems.

The near-coincident-pair technique has these features:

- It uses two directional mics angled apart and spaced a few inches apart.
- Level and time differences between channels produce the stereo effect.
- Images are sharp.
- Stereo spread tends to be accurate.
- It provides a greater sense of "air" and depth than coincident methods.

Mounting Hardware

With coincident and near-coincident techniques, the microphones should be rigidly mounted with respect to one another, so that they can be moved as a unit without disturbing their arrangement. A device for this purpose is called a *stereo microphone adapter* or *stereo bar.* It mounts two microphones on a single stand, and microphone angling and spacing are adjustable. A number of these devices are listed in Chapter 16.

Microphone Requirements

The sound source dictates the requirements of the recording microphones. Most acoustic instruments produce frequencies from about 40 Hz (string bass and bass drum) to about 20,000 Hz (cymbals, castanets, triangles). A microphone with uniform response between these frequency limits will do full justice to the music.

The highest octave from 10 to 20 kHz adds transparency, air, and realism to the recording. You may need to roll off (filter out) frequencies below 80 Hz to eliminate rumble from trucks and air conditioning, unless you want to record organ or bass-drum fundamentals.

Sound from an orchestra or band approaches each microphone from a broad range of angles. To reproduce all the instruments' timbres equally well, the microphone should have a broad, flat response at all angles of incidence within at least 90°; that is, the polar pattern should be uniform with frequency. Microphones with small-diameter diaphragms usually meet this requirement best. (Note that some microphones have small diaphragms inside large housings.)

If you're forced to record at a great distance, a frequency response elevated up to 4 dB above 4 kHz might sound more natural than a flat response. Another benefit of a rising high end is that you can roll it off in postproduction, reducing analog tape hiss. Since classical music covers a wide dynamic range (up to 80 dB), the recording microphones should have very low noise and distortion. In distant-miking applications, the sensitivity should be high to override mixer noise.

For sharp imaging, the microphone pair should be well matched in frequency response, phase response, and polar pattern.

We investigated several microphone arrangements for recording in stereo. Each has its advantages and disadvantages. Which method you choose depends on the sonic compromises you are willing to make.

References

B. Bartlett. "Stereo Microphone Technique." *db* 13, no. 12 (December 1979), pp. 34–46.

J. Eargle. "Stereo Microphone Techniques," Chapter 10. In *The Microphone Handbook.* Plainview, NY: Elar Publishing, 1981.

David Griesinger. "Spaciousness and Localization in Listening Rooms and Their Effects on the Recording Technique." *Journal of the Audio Engineering Society* 34, no. 4 (April 1986), pp. 255–268.

S. Lipshitz. "Stereo Microphone Techniques: Are the Purists Wrong?" *Journal of the Audio Engineering Society* 34, no. 9 (September 1986), pp. 716–744.

R. Streicher and W. Dooley. "Basic Stereo Microphone Perspectives—A Review." *Journal of the Audio Engineering Society* 33, nos. 7–8 (July–August 1984), pp. 548–556.

9

STEREO IMAGING THEORY

A sound system with good stereo imaging can form apparent sources of sound, such as reproduced musical instruments, in well-defined locations, usually between a pair of loudspeakers placed in front of the listener. These apparent sound sources are called *images.*

This chapter explains terms related to stereo imaging, how we localize real sound sources, how we localize images, and how microphone placement controls image location.

You can use stereo microphone techniques without reading this chapter. However, if you want to deepen your understanding of what's going on or develop your own stereo array, it's worthwhile to study the theory and simple math in this chapter.

Definitions

First, I define several terms related to stereo imaging. *Fusion* refers to the synthesis of a single apparent source of sound (an image or "phantom image") from two or more real sound sources (such as loudspeakers).

The *location* of an image is its angular position relative to a point straight ahead of a listener, or its position relative to the loudspeakers. This is shown in Figure 9–1. An aim of high fidelity is to reproduce the images in the locations intended by the recording engineer or producer. In some productions, usually classical music recordings, a goal of the recording engineer or producer is to place the images in the same relative locations as the instruments were during the live performance.

Stereo spread or *stage width* (Figure 9–2) is the distance between the extreme left and right images of a reproduced ensemble of instruments.

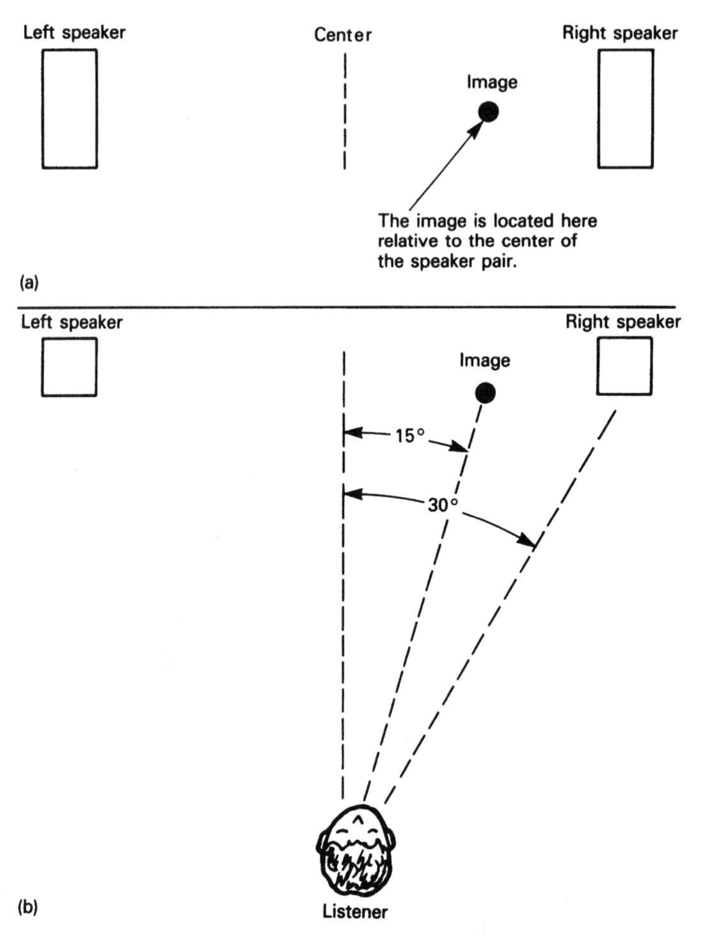

Figure 9–1 Example of image location: (a) listener's view; (b) top view.

The stereo spread is wide if the ensemble appears to spread all the way between a pair of loudspeakers. The spread is narrow if the reproduced ensemble occupies only a small space between the speakers. Sometimes, the reproduced reverberation or ambience spreads from speaker to speaker even when the reproduced ensemble width is narrow.

Image *focus* or *size* (Figure 9–3) refers to the degree of fusion of an image, its positional definition. A sharply focused image is described as

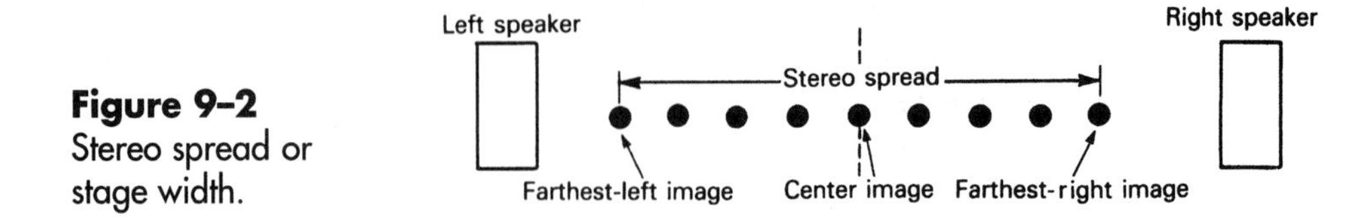

Figure 9–2 Stereo spread or stage width.

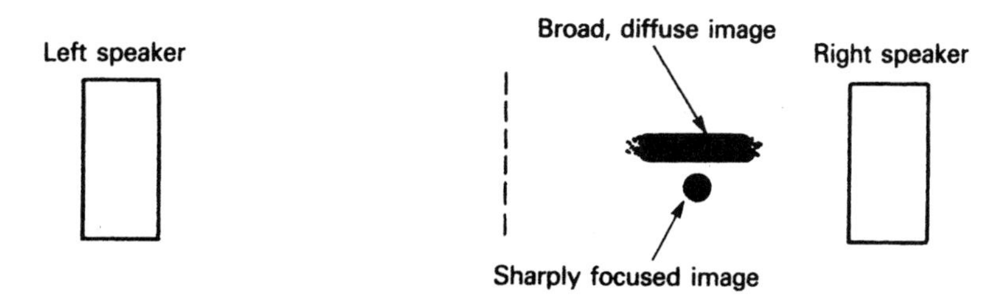

Figure 9–3 Image focus or size (listener's perception).

being pinpointed, precise, narrow, sharp, resolved, well-defined, or easy to localize. A poorly focused image is hard to localize; it is spread, broad, smeared, vague, diffuse. A natural image is focused to the same degree as the real instrument being reproduced.

Depth is the apparent distance of an image from a listener, the sense of closeness and distance of various instruments.

Elevation is an image displacement in height above the speaker plane.

Image *movement* is a reproduction of the movement of the sound source, if any. The image should not move unaccountably.

Localization is the ability of a listener to tell the direction of a sound. It also is the relation between interchannel or interaural differences and perceived image location.

How We Localize Real Sound Sources

The human hearing system uses the direct sound and early reflections to localize a sound source. The direct sound and reflections within about 2 milliseconds contribute to localization (Wallach, Newman, and Rosenzweig, 1973; Bartlett, 1979). Reflections occurring up to 5–35 milliseconds after the direct sound influence image broadening Gardner, 1973). Distance or depth cues are conveyed by early reflections (less than 33 milliseconds after the direct sound). Echoes delayed more than about 5–50 milliseconds (depending on program material) do not fuse in time with the early sound but contribute to the perceived tonal spectrum (Carterette and Friedman, 1978, pp. 62, 210).

Imagine a sound source and a listener. Let's say that the source is in front of the listener and to the left of center (as in Figure 9–4). Sound travels a longer distance to the right ear than to the left ear, so the sound arrives at the right ear after it arrives at the left ear. In other words, the right-ear signal is delayed relative to the left-ear signal. Every source location produces a unique arrival-time difference between ears (Vanderlyn, 1979).

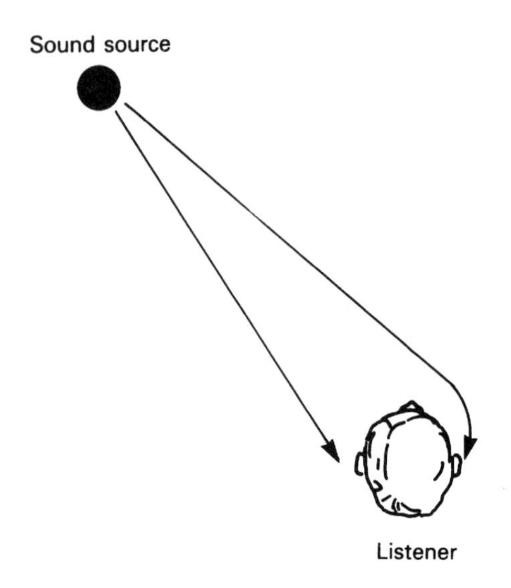

Figure 9–4
Sound traveling from a source to a listener's ears.

In addition, the head acts as an obstacle to sounds above about 1000 Hz. High frequencies are shadowed by the head, so a different spectrum (amplitude versus frequency) appears at each ear (Shaw, 1974; Mehrgardt and Mellert). Every source location produces a unique spectral difference between ears (Figure 9–5).

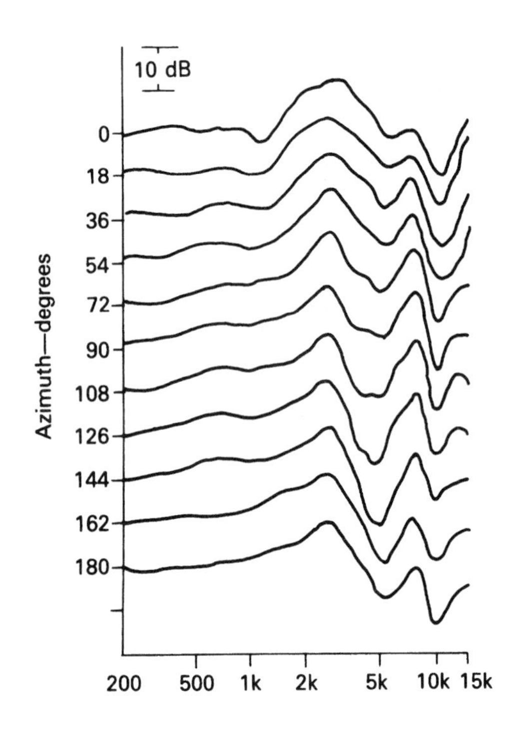

Figure 9–5
Frequency response of the ear at different azimuth angles: 0° is straight ahead; 90° is to the side of the ear being measured; 180° is behind the head (after Mehrgardt and Mellert).

We have learned to associate certain interaural differences with specific directions of the sound source. When presented with a new source location, we match what we hear with a memorized pattern of a similar situation to determine direction (Rumsey, 1989, p. 6).

As stated before, an important localization cue is the interaural arrival-time difference of the signal envelope. We perceive this difference at any change in the sound—a transient, a pause, or a change in timbre. For this reason, we localize transients more easily than continuous sounds (Rumsey, 1989, p. 3). The time difference between ear signals also can be considered a phase difference between sound waves arriving at the ears (Figure 9–6).This phase shift rises with frequency.

When sound waves from a real source strike a listener's head, a different spectrum of amplitude and phase appears at each ear. These interaural differences are translated by the brain into a perceived direction of the sound source. Every direction is associated with a different set of interaural differences.

The ears make use of interaural phase differences to localize sounds between about 100 and 700 Hz. Frequencies below about 100 Hz are not localized (making "subwoofer/satellite" speaker systems feasible; Harvey and Schroeder, 1961). Above about 1500 Hz, amplitude differences between ears contribute to localization. Between about 700 and 1500 Hz, both phase and amplitude differences are used to tell the direction of a sound (Eargle, 1976, Chapters 2 and 3; Cooper and Bauck, 1980).

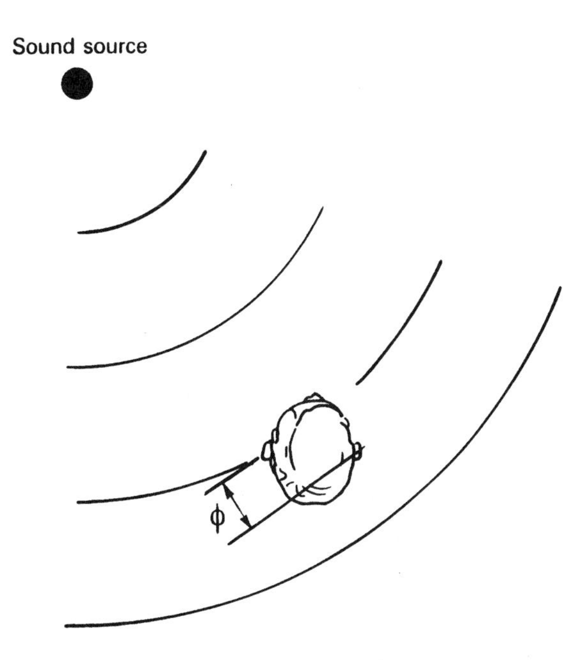

Figure 9–6
Phase shift ϕ between sound waves at the ears.

Small movements of the head change the arrival-time difference at the ears. The brain uses this information as another cue for source location (Rumsey, 1989, p. 4), especially for distance and front/back discrimination.

The outer ears (pinnae) play a part as well (Gardner and Gardner, 1973). In each pinna, sound reflections from various ridges combine with the direct sound, causing phase cancellation-frequency notches in the perceived spectrum. The positions of the notches in the spectrum vary with the source height. We perceive these notch patterns not as a tonal coloration but as height information. Also, we can discriminate sounds in front from those in back because of the pinnae's shadowing effect at high frequencies.

Some of the cues used by the ears can be omitted without destroying localization accuracy if other cues are still present.

How We Localize Images Between Speakers

Now that we discussed how we localize real sound sources, let's look at how we localize their images reproduced over loudspeakers. Imagine that you're sitting between two stereo speakers, as in Figure 9–7. If you feed a musical signal equally to both channels in the same polarity, you'll perceive an image between the two speakers. Normally, you'll hear a single synthetic source, rather than two separate loudspeaker sources.

Each ear hears both speakers. For example, the left ear hears the left speaker signal then, after a short delay due to the longer travel path, hears the right speaker signal. At each ear, the signals from both speakers sum or add together vectorially to produce a resultant signal.

Suppose that we make the signal louder in one speaker. That is, we create a level difference between the speakers. Surprisingly, this causes an arrival-time difference at the ears (Rumsey, 1989, p. 8). This is a result of the phasor addition of both speaker signals at each ear.

Remember to distinguish interchannel differences (between speaker channels) from interaural differences (between ears). An interchannel level difference does not appear as an interaural level difference but rather as an interaural time difference.

We can use this speaker-generated interaural time difference to place images. Here is how: Suppose we want to place an image 15° to one side. A real sound source 15° to one side produces an interaural time difference of 0.13 millisecond. If we can make the speakers produce an interaural time difference of 0.13 millisecond, we'll hear the image 15° to one side. We can fool the hearing system into believing there's a real source at that angle. This occurs when the speakers differ in level by a certain amount.

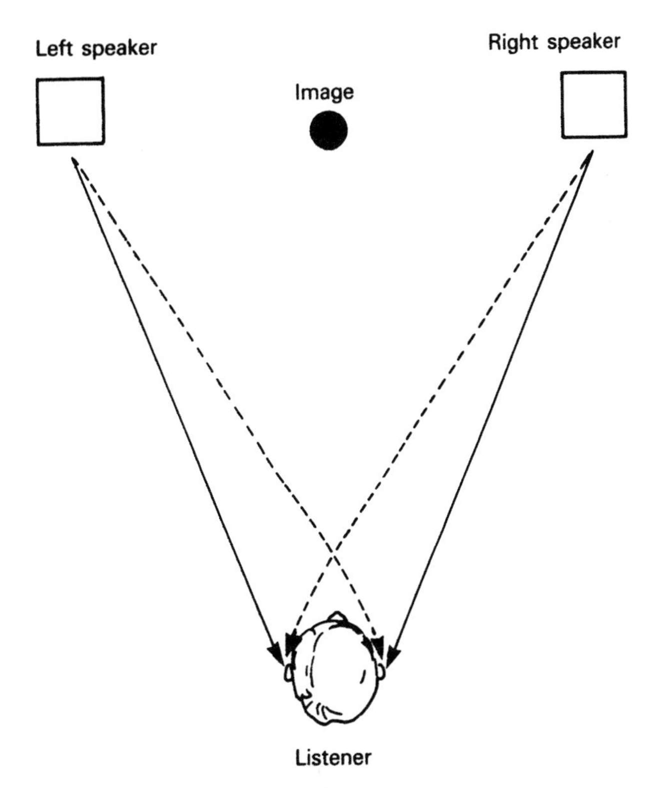

Figure 9–7 Two ears receiving signals from two speakers.

The polarity of the two channels affects localization as well. To explain polarity, if the signals sent to two speaker channels are in polarity, they are in phase at all frequencies; both go positive in voltage at the same time. If the signals are out of polarity, they are 180° out of phase at all frequencies. One channel's signal goes positive when the other channel's signal goes negative. Opposite polarity signals sometimes are incorrectly referred to as being *out of phase.*

If the signals are in opposite polarity between channels and equal level in both channels, the resulting image has a diffuse, directionless quality and cannot be localized. If the signals are in opposite polarity and at a higher level in one channel than the other, the image often appears outside the bounds of the speaker pair. You'd hear an image left of the left speaker or right of the right speaker (Eargle, 1976).

Opposite polarity can occur in several ways. Two microphones are of opposite polarity if the wires to connector pins 2 and 3 are reversed in one microphone. Two speakers are of opposite polarity if the speaker-cable leads are reversed at one speaker. A single microphone might have different parts of its polar pattern in opposite polarity. For example, the rear lobe of a bidirectional pattern is opposite in polarity to the front lobe. If sound from a particular direction reaches the front lobe of the left-channel mic and the rear lobe of the right-channel mic, the two channels will be of opposite polarity. The resulting image of that sound source will either be diffuse or outside the speaker pair.

Requirements for Natural Imaging over Loudspeakers

To the extent that a sound recording and reproducing system can duplicate the interaural differences produced by a real source, the resultant image will be accurately localized. In other words, when reproduced sounds reaching a listener's ears have amplitude and phase differences corresponding to those of a real sound source at some particular angle, the listener perceives a well-fused, naturally focused image at that same angle. Conversely, when unnatural amplitude and phase relations are produced, the image appears relatively diffuse rather than sharp and is harder to localize accurately (Cooper and Bauck, 1980).

The required interaural differences for realistic imaging can be produced by certain interchannel differences. Placing an image in a precise location requires a particular amplitude difference versus frequency and phase difference versus frequency between channels. These have been calculated by Cooper and Bauck (1980) for several image angles. Gerzon (1980), Nakabayashi (1975), and Koshigoe and Takahashi (1976) have calculated the interaural or interchannel differences required to produce any image direction at a single frequency.

Figure 9–8, for example, shows the interchannel differences required to place an image at 15° to the left of center when the speakers are placed ±30° in front of the listener (Cooper and Bauck, 1980).

As Figure 9–8 shows, the interchannel differences required for natural imaging vary with frequency. Specifically, Cooper and Bauck (1980) indicate that interchannel amplitude differences are needed below approximately 1700 Hz and interchannel time differences are needed above that frequency (Cooper, 1987). Specifically,

- At low frequencies, the amplitude difference needed for a 15° image angle is about 10 dB.
- Between 1.7 kHz and 5 kHz, the amplitude difference goes to approximately 0 dB.

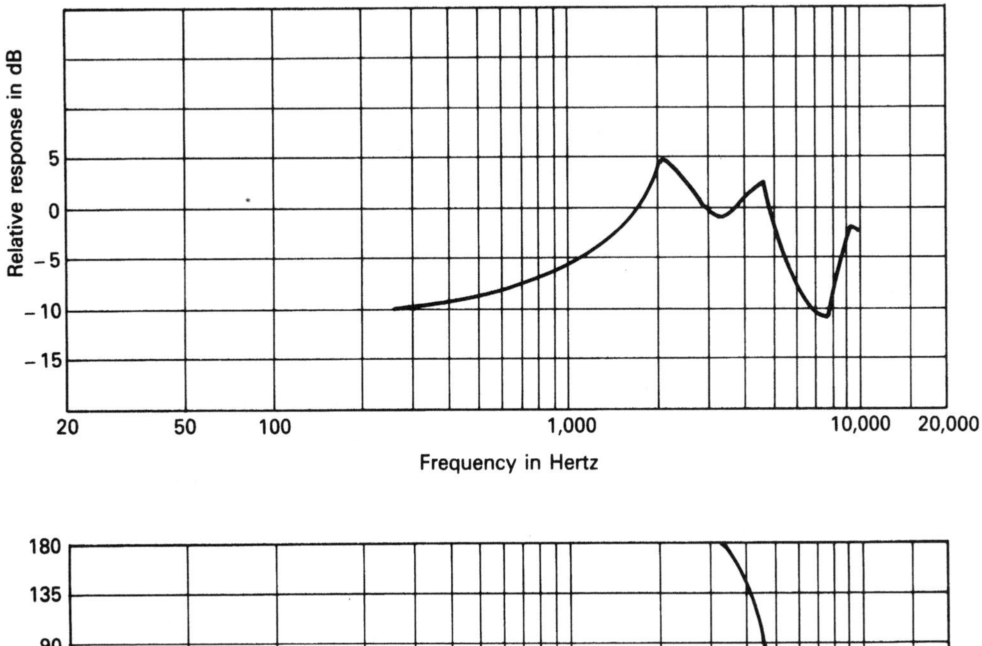

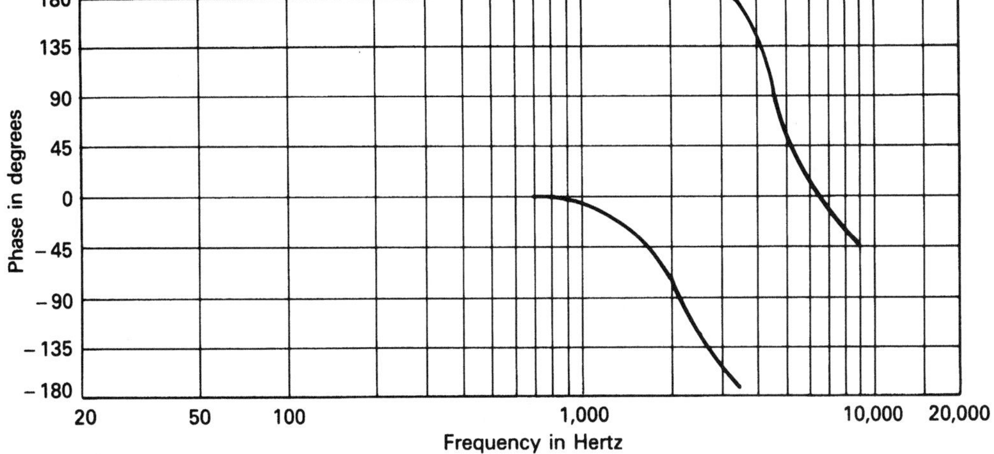

Figure 9–8 Amplitude (top) and phase (bottom) of right channel relative to left channel, for image location 15° to the left of center when speakers are ±30° in front of listener.

- Above 1.7 kHz, the phase difference corresponds to a group delay (interchannel time difference) of about 0.547 millisecond, or 7.39 inches for a hypothetical spacing between microphones used for stereo recording.

This theory is based on the "shadowing" of sound traveling around a sphere. The description given here simplifies the complex requirements, but it conveys the basic idea. Cooper notes that "moderate deviations from these specifications might not lead to noticeable auditory distress or faulty imaging."

The Cooper-Bauck criteria can be met by recording with a dummy head whose signals are specially processed. A dummy head used for binaural recording is a modeled head with a flush-mounted microphone in each ear. Time and spectral differences between channels create the stereo images. (Spectral differences are amplitude differences that vary with frequency.)

Although a dummy-head binaural recording can provide excellent imaging over headphones, it produces poor localization over loudspeakers at low frequencies (Huggonet and Jouhaneau, 1987, Figure 13, p. 16) unless spatial equalization (a shuffler circuit) is used (Griesinger, 1989). Spatial equalization boosts the low frequencies in the difference (L – R) signal.

Binaural recording can produce images surrounding a listener wearing headphones but only frontal images over loudspeakers, unless a transaural converter is used. A transaural converter is an electronic device that converts binaural signals (for headphone playback) into stereo signals (for loudspeaker playback). Transaural stereo is a method of surround-sound reproduction using a dummy head for binaural recording, processed electronically to remove head-related crosstalk when the recording is heard over two loudspeakers (Eargle, 1976, pp. 122–123; Bauer, 1961; Schroeder and Atal, 1963; Damaske, 1971; Mori et al., 1979; Sakamoto et al., 1978, 1981, 1982; Moller, 1989; Cooper and Bauck, 1989). (More on this in Chapter 12.)

Cooper recommends that, for natural imaging, the speakers' interchannel differences be controlled so that their signals sum at the ears to produce the correct interaural differences. According to Theile (1987), Cooper's theory (based on summing localization) is in error because it applies only to sine waves and may not apply to broadband spectral effects. He proposes a different theory of localization, the association model. This theory suggests that, when listening to two stereo loudspeakers, we ignore our interaural differences and instead use the speakers' interchannel differences to localize images.

The interchannel differences needed for best stereo, Theile says, are head related. The ideal stereo-miking technique would use, perhaps, two ear-spaced microphones flush mounted in a head-sized sphere and equalized for flat subjective response. This would produce interchannel spectral and time differences that, Theile claims, are optimum for stereo. The interchannel differences—time differences at low frequencies, amplitude differences at high frequencies—are the opposite of Cooper's requirements for natural stereo imaging. Time will tell which theory is closer to the truth.

Currently Used Image-Localization Mechanisms

The ear can be fooled into hearing reasonably sharp images between speakers by less sophisticated signal processing. Simple amplitude or time differences between channels, constant with frequency, can produce localizable images. Bartlett (1979, pp. 38, 40), Masden (1957), Dutton (1962), Cabot (1977), Williams (1987), Blauert (1983, pp. 206–207), and Rumsey (1989) give test results showing image location as a function of interchannel amplitude or time differences. Bartlett's results are shown later in this chapter.

For example, given a speech signal, if the left channel is 7.5 dB louder than the right channel, an image will appear at approximately 15° to the left of center when the speakers are placed ±30° in front of the listener. A delay in the right channel of about 0.5 millisecond will accomplish the same thing, although image locations produced solely by time differences are relatively vague and hard to localize.

Griesinger notes that pure interchannel time differences do not localize lowpass-filtered male speech below 500 Hz over loudspeakers. Amplitude (level) differences are needed to localize low-frequency sounds. Either amplitude or time differences can localize high-frequency sounds (Griesinger, 1987). However, Blauert (1983) gives evidence by Wendt that interchannel time differences do cause localization at 327 Hz.

The interchannel differences produced in current stereo recordings are just simple approximations of what is required. Current practice uses interchannel amplitude or time differences, or both, to locate images. These differences are constant with frequency (except in baffled omnis). Still, reasonably sharp images are produced. Let's look at exactly how these differences localize images (Bartlett, 1979) .

Localization by Amplitude Differences

The location of images between two loudspeakers depends in part on the signal amplitude differences between the loudspeakers. Suppose a speech signal is sent to two stereo loudspeakers, with the signal to each speaker identical except for an amplitude (level) difference (as shown in Figure 9–9). We create an amplitude difference by inserting an attenuator in one channel.

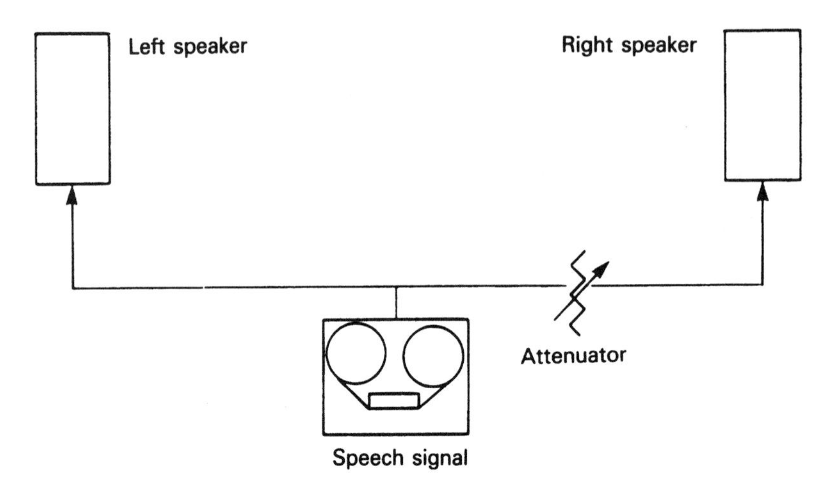

Figure 9–9 Sending a speech signal to two stereo loudspeakers with attenuation in one channel.

Figure 9–10 shows the approximate sound image location between speakers versus the amplitude difference between channels, in decibels. A 0 dB difference (equal level from each speaker) makes the image of the sound source appear in the center, midway between the speakers. Increasing the difference places the image farther away from the center. A difference of 15–20 dB makes the image appear at only one speaker.

The information in this figure is based on carefully controlled listening tests. The data are the average of the responses of 10 trained listeners. They auditioned a pair of signal-aligned, high-quality loudspeakers several feet from the walls in a "typical" listening room, while sitting centered between the speakers at a 60° listening angle.

How can we create this effect with a stereo microphone array? Suppose two cardioid microphones are crossed at 90° to each other, with the grille of one microphone directly above the other (Figure 9–11). The microphones are angled 45° to the left and right of the center of the orchestra. Sounds arriving from the center of the orchestra will be picked up equally by both microphones. During playback, both speakers will be at equal levels, and consequently, a center image is produced.

Figure 9–10 Stereo-image location versus amplitude difference between channels, in dB (listener's perception).

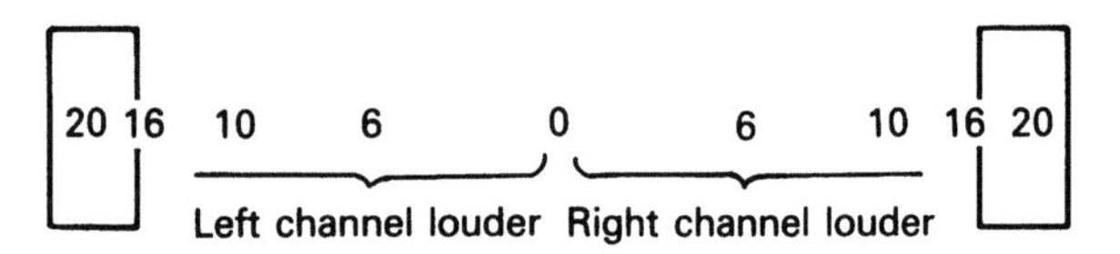

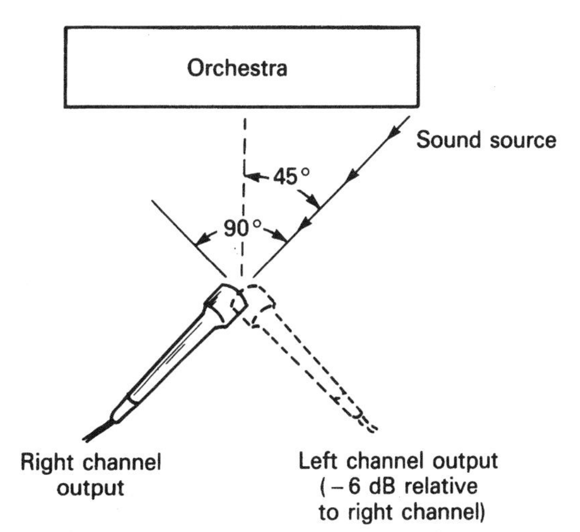

Figure 9–11
Cardioids crossed at 90°, picking up a source at one end of an orchestra.

Suppose that the extreme right side of the orchestra is 45° off center, from the viewpoint of the microphone pair. Sounds arriving from the extreme right side of the orchestra approach the right-aiming microphone on axis, but they will approach the left-aiming microphone at 90° off axis (as shown in Figure 9–11). A cardioid polar pattern has a 6 dB lower level at 90° off axis than it has on axis. So, the extreme right sound source will produce a 6 dB lower output from the left microphone than from the right microphone.

Thus, we have a 6 dB amplitude difference between channels. According to Figure 9–10, the image of the extreme right side of the orchestra will now be reproduced right of center. Instruments in between the center and the right side of the orchestra will be reproduced somewhere between the 0 and 6 dB points.

If we angle the microphones farther apart, for example, 135°, the difference produced between channels for the same source is around 10 dB. As a result, the right-side stereo image will appear farther to the right than it did with 90° angling. (Note that it is not necessary to aim the microphones exactly at the left and right sides of the ensemble.)

The farther to one side a sound source is, the greater the amplitude difference between channels it produces and, so, the father from center is its reproduced sound image.

Localization by Time Differences

Phantom-image location also depends on the signal time differences between channels. Suppose we send the same speech signal to two speakers at equal levels but with one channel delayed (as in Figure 9–12).

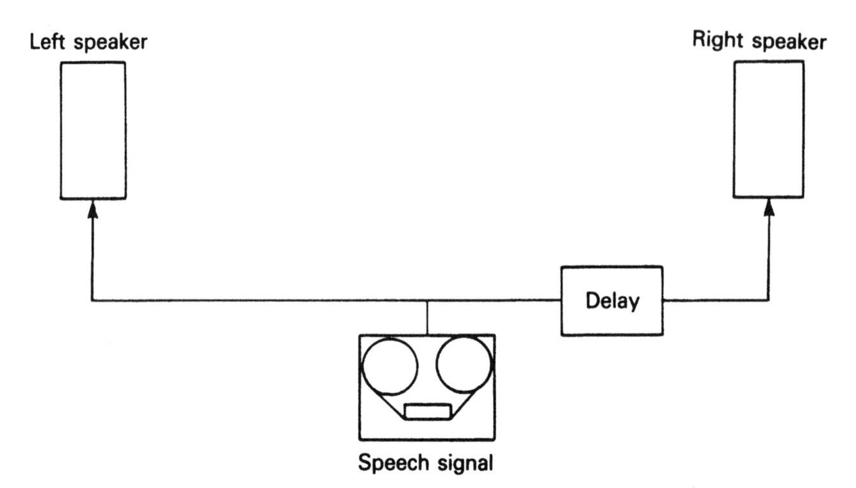

Figure 9–12 Sending a speech signal to two speakers with one channel delayed.

Figure 9–13 shows the approximate sound image location between speakers, with various time differences between channels, in milliseconds. A 0 millisecond difference (no time difference between speaker channels) makes the image appear in the center. As the time difference increases, the phantom image appears farther off center. A 1.5 millisecond difference or delay is sufficient to place the image at only one speaker.

Spacing two microphones apart horizontally—even by a few inches—produces a time difference between channels for off-center sources. A sound arriving from the right side of the orchestra will reach the right microphone first, simply because it is closer to the sound source (as in Figure 9–14). For example, if the sound source is 45° to the right and the microphones are 8 inches apart, the time difference produced between channels for this source is about 0.4 millisecond. For the same source, a 20 inch spacing between microphones produces a 1.5 millisecond time difference between channels, placing the reproduced sound image at one speaker.

With spaced-pair microphones, the farther a sound source is from the center of the orchestra, the greater the time difference between channels and, so, the farther from center is its reproduced sound image.

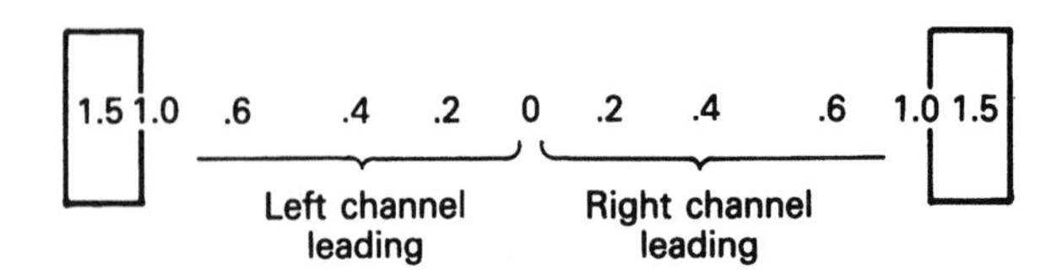

Figure 9–13 Approximate image location versus time difference between channels, in milliseconds (listener's perception).

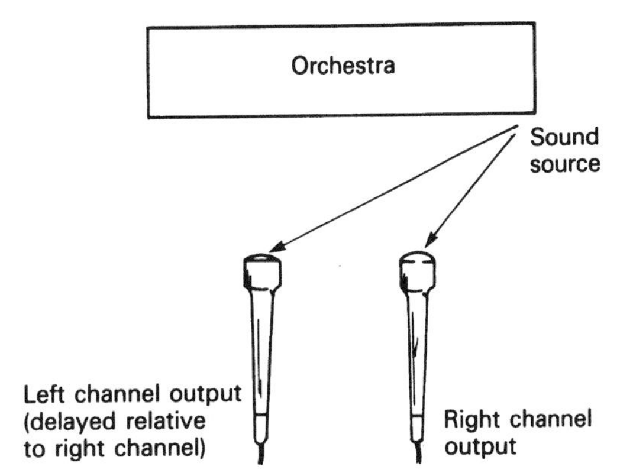

Figure 9–14
Microphones spaced apart, picking up a source at one end of an orchestra.

Localization by Amplitude and Time Differences

Phantom images also can be localized by a combination of amplitude and time differences. Suppose 90° angled cardioid microphones are spaced 8 inches apart (as in Figure 9–15). A sound source 45° to the right will produce a 6 dB level difference between channels and a 0.4 millisecond difference between channels. The image shift of the 6 dB level difference adds to the image shift of the 0.4 millisecond difference to place the sound image at the right speaker. Certain other combinations of angling and spacing accomplish the same thing.

Summary

If a speech signal is recorded on two channels, its reproduced sound image will appear at only one speaker when the signal is 15 to 20 dB lower in one

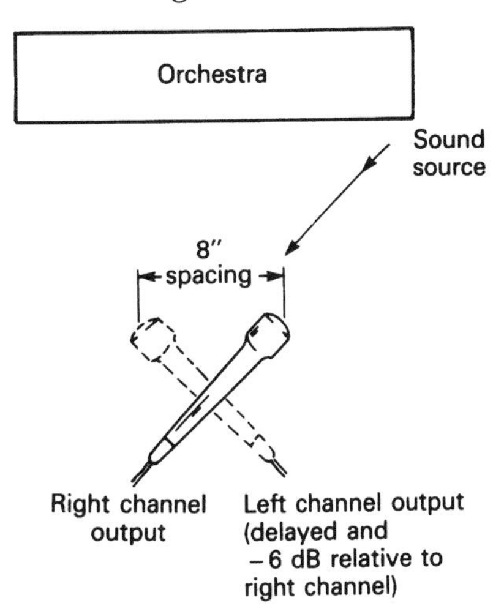

Figure 9–15
Cardioids angled 90° and spaced 8 inches, picking up a source at one end of an orchestra.

channel, or the signal is delayed 1.5 milliseconds in one channel, or the signal in one channel is lower in level and delayed by a certain amount.

When amplitude and time differences are combined to place images, the sharpest imaging occurs when the channel that is lower in level is also the channel that is delayed. If the higher-level channel is delayed, image confusion results due to the conflicting time and amplitude cues.

We have seen that angling directional microphones (coincident placement) produces amplitude differences between channels. Spacing microphones (spaced-pair placement) produces time differences between channels. Angling and spacing directional microphones (near-coincident placement) produces both amplitude and time differences between channels. These differences localize the reproduced sound image between a pair of loudspeakers.

Predicting Image Locations

Suppose you have a pair of microphones for stereo recording. Given their polar pattern, angling, and spacing, you can predict the interchannel amplitude and time differences for any source angle. Hence, you can predict the localization of any stereo microphone array.

This prediction assumes that the microphones have ideal polar patterns, and that these patterns do not vary with frequency. It's an unrealistic assumption, but the prediction agrees well with listening tests.

The amplitude difference between channels in dB is given by

$$\Delta dB = 20 \log \left[\frac{a + b\cos((\theta_m/2) - \theta_s)}{a + b\cos((\theta_m/2) + \theta_s)} \right] \tag{1}$$

where

ΔdB = amplitude difference between channels, in dB;

$a + b\cos(\theta)$ = polar equation for the microphone;

Omnidirectional:	$a = 1$	$b = 0$
Bidirectional:	$a = 0$	$b = 1$
Cardioid:	$a = 0.5$	$b = 0.5$
Supercardioid:	$a = 0.366$	$b = 0.634$
Hypercardioid:	$a = 0.25$	$b = 0.75$

θ_m = angle between microphone axes, in degrees;

θ_s = source angle (how far off center the sound source is), in degrees.

These variables are shown in Figure 9–16.

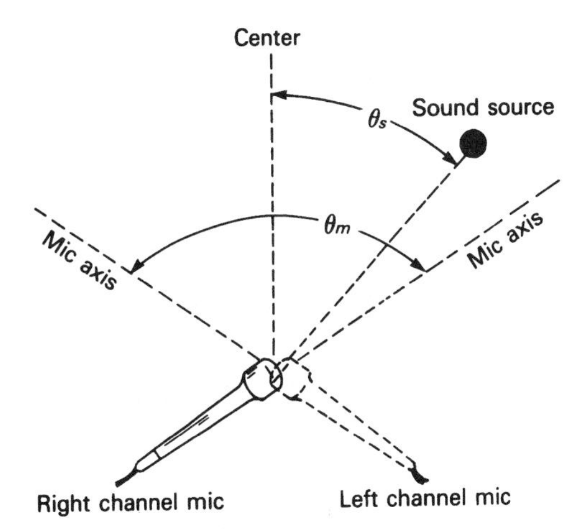

Figure 9–16
Microphone angle (θ_m), source angle (θ_s).

The time difference between channels is given by

$$\Delta T = \frac{\sqrt{D^2 + [(S/2) + D \tan \theta_s)]^2} - \sqrt{D^2 + [(S/2) - D \tan \theta_s)]^2}}{C} \qquad (2)$$

where

ΔT = time difference between channels, in seconds;
D = distance from the source to the line connecting the microphones, in feet;
S = spacing between microphones, in feet;
θ_s = source angle (how far off center the sound source is), in degrees;
C = speed of sound (1130 feet per second).

These variables are shown in Figure 9–17.

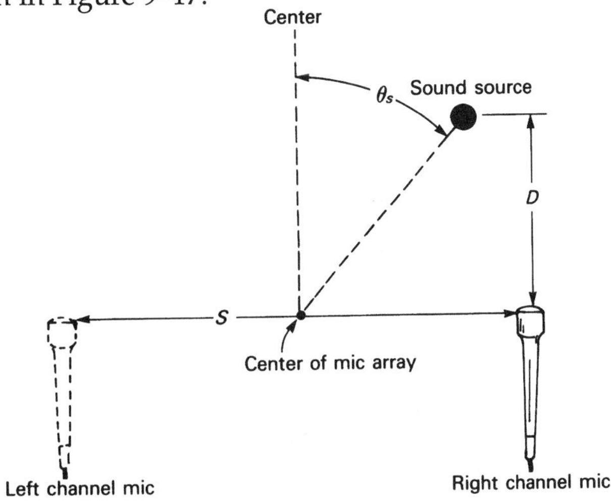

Figure 9–17
Source angle (θ_s), mic-to-source distance (D), and mic spacing (S).

For near-coincident microphone spacing of a few inches, the equation can be simplified to this:

$$\Delta T = \frac{S \sin \theta_s}{C} \qquad (3)$$

ΔT = time difference between channels, in seconds;
S = microphone spacing, in inches;
θ_s = source angle, in degrees;
C = speed of sound (13,560 inches per second).

These variables are shown in Figure 9–18.

Let's consider an example. If you angle two cardioid microphones 135° apart and the source angle is 60° (as in Figure 9–19), the dB difference produced between channels for that source is calculated as follows.

For a cardioid, $a = 0.5$ and $b = 0.5$ (from the list following equation (1)). The angle between microphone axes, θ_m, is 135°; the source angle, θ_s, is 60°. That is, the sound source is 60° off center. So the amplitude difference between channels, using equation (1), is

$$\Delta \text{dB} = 20 \log \left[\frac{0.5 + 0.5 \cos((135°/2) - 60°)}{0.5 + 0.5 \cos((135°/2) + 60°)} \right]$$

$$= 14 \text{ dB amplitude difference between channels}$$

So, according to Figure 9–10, that sound source will be reproduced nearly all the way at one speaker.

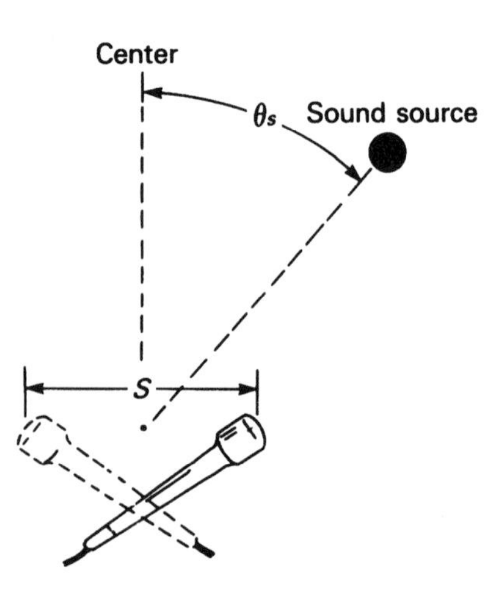

Figure 9–18
Mic spacing (S) and source angle (θ_s).

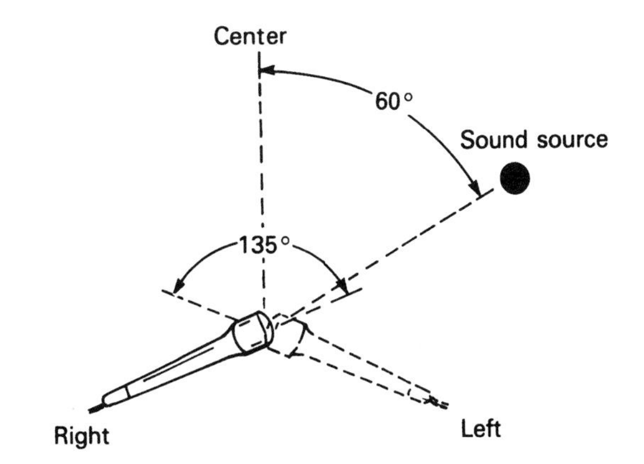

Figure 9–19
Cardioids angled 135° apart, with a 60° source angle.

Here is another example. If you place two omnidirectional microphones 10 inches apart and the sound source is 45° off center, what is the time difference between channels? (Refer to Figure 9–20.)

Microphone spacing, S, is 10 inches; source angle, θ_s, is 45°. By equation (3),

$$\Delta T = \frac{10 \sin 45°}{13{,}560}$$

= 0.52 millisecond time difference between channels

So, according to Figure 9–10, that sound source will be reproduced about halfway off center.

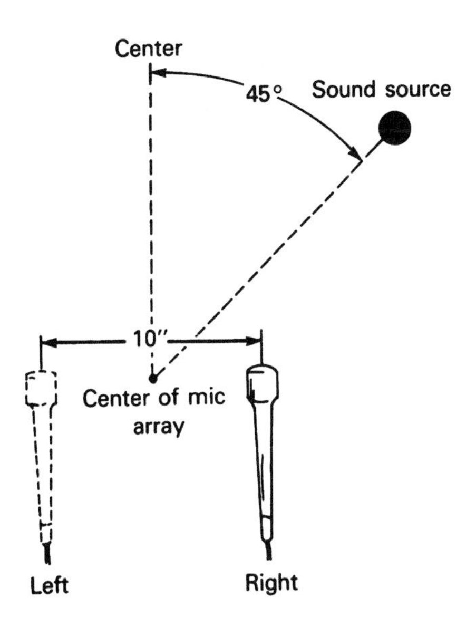

Figure 9–20
Omnis spaced 10 inches apart, with a 45° source angle.

Choosing Angling and Spacing

Many combinations of microphone angling and spacing are used to place the images of the ends of the orchestra at the right and left speaker. In other words, there are many ways to achieve a full stereo spread. You can use a narrow spacing and a wide angle or a wide spacing and a narrow angle—whatever works. The particular angle and spacing you use is not sacred. Many do not realize this, and rely on a fixed angle or spacing, such as the ORTF system (110°, 17 cm). That is a good place to start, but if the reproduced stage width is too narrow, there's no harm in increasing the angle or spacing slightly.

If the center instruments are too loud, you can angle the mics farther apart while decreasing the spacing so that the reproduced stage width is unchanged. In this way, you can control the loudness of the center image to improve the balance.

To reduce pickup of early reflections from the stage floor and walls, (1) increase angling, (2) decrease spacing, and (3) place the mics closer to the ensemble. This works as follows:

1. Angling the mics farther apart softens the center instruments.
2. Decreasing the spacing between mics maintains the original reproduced stage width.
3. Since center instruments are quieter, you can place the mics closer to the ensemble and still achieve a good balance.
4. Since the mics are closer, the ratio of reflected sound to direct sound is decreased. You can add distant mics or artificial reverberation for the desired amount of hall ambience.

In general, a combination of angling and spacing (intensity and time differences) gives more accurate localization and sharper imaging than intensity or time differences alone (Griesinger, 1987).

Angling the mics farther apart increases the ratio of reverberation in the recording, which makes the orchestra sound farther away. Spacing the mics farther apart does not change the sense of distance, but it degrades the sharpness of the images.

Spaciousness and Spatial Equalization

The information in this section is from Griesinger's (1987) paper, "New Perspectives on Coincident and Semi-Coincident Microphone Arrays."

The spaciousness of a microphone array is the ratio of L – R energy to L + R energy in the reflected sound. Ideally, this ratio should be equal to or greater than 1. In other words, the sum and difference energy are equal. Spaciousness implies a low correlation between channels of the reflected sound.

Some microphone arrays with good spaciousness (a value of 1) are the spaced pair, the Blumlein pair (figure eights crossed at 90°), the MS array with a cardioid mid pattern and a 1:1 M/S ratio, and coincident hypercardioids angled 109° apart.

Spatial equalization or shuffling is a low-frequency shelving boost of difference (L – R) signals and a complementary low-frequency shelving cut of sum (L + R) signals. This has two benefits:

1. It increases spaciousness, so that coincident and near-coincident arrays can sound as spacious as spaced arrays.
2. It aligns the low-frequency and high-frequency components of the sound images, which results in sharper image focus.

You can build a spatial equalizer as shown in Griesinger's paper. Or use an MS technique and boost the low frequencies in the L – R or side signal, and cut the low frequencies in the L + R or mid signal. The required boost or cut depends on the mic array, but a typical value is 4–6 dB shelving below 400 Hz. Excessive boost can split off-center images, with bass and treble at different positions. The correction should be done to the array before it is mixed with other mics.

Gerzon (1987) points out that the sum and difference channels should be phase compensated (with matched, nonminimum phase responses of the two filters), as suggested by Vanderlyn (1957). Gerzon notes that spatial equalization is best applied to stereo microphone techniques not having a large antiphase reverberation component at low frequencies, such as coincident or near-coincident cardioids. With the stereosonic technique of crossed figure eights, antiphase components tend to become excessive. He suggests a 2.4 dB cut in the sum (L + R) signal and a 5.6 dB boost in the difference (L – R) signal for better bass response.

Griesinger states,

> Spatial equalization can be very helpful in coincident and semi-coincident techniques [especially when listening is done in small rooms]. Since the strongest localization information comes from the high frequencies, microphone patterns and angles can be chosen which give an accurate spread to the images at high frequencies. Spatial equalization can then be used to raise the spaciousness at low frequencies.

Alan Blumlein devised the first shuffler, revealed in his 1933 patent. He used it along with two omni mic capsules spaced apart the width of a human head. The shuffler differenced the two channels (added them in opposite polarity). When two omnis are added in opposite polarity, the result is a single bidirectional pattern aiming left and right. Blumlein used this pattern as the side pattern in an MS pair (Lipschitz, 1990).

The frequency response of the synthesized bidirectional pattern is weak in the bass: It falls 6 dB/octave as frequency decreases. So Blumlein's shuffler circuit also included a first-order low-frequency boost (below 700 Hz) to compensate.

The shuffler converts phase differences into intensity differences. The farther off center the sound source is, the greater the phase difference between the spaced mics. And the greater the phase difference, the greater is the intensity difference between channels created by the shuffler.

References

B. Bartlett. "Stereo Microphone Technique." *db* 13, no. 12 (December 1979), pp. 34–46.

B. Bauer. "Stereophonic Earphones and Binaural Loudspeakers." *Journal of the Audio Engineering Society* 9, no. 2 (April 1961), pp. 148–151.

J. Blauert. *Spatial Hearing*. Cambridge, MA: MIT Press, 1983.

R. Cabot. "Sound Localization in Two and Four Channel Systems: A Comparison of Phantom Image Prediction Equations and Experimental Data." Preprint No. 1295 (J3), paper presented at the Audio Engineering Society 58th convention, November 4–7, 1977, New York.

E. Carterette and M. Friedman. *Handbook of Perception,* vol. 4, *Hearing.* New York: Academic Press, 1978.

D. H. Cooper. "Problems with Shadowless Stereo Theory: Asymptotic Spectral Status." *Journal of the Audio Engineering Society* 35, no. 9 (September 1987), p. 638.

D. Cooper and J. Bauck. "On Acoustical Specification of Natural Stereo Imaging." Preprint No. 1616 (X3), paper presented at the Audio Engineering Society 65th convention, February 25–28, 1980, London.

D. Cooper and J. Bauck. "Prospects for Transaural Recording." *Journal of the Audio Engineering Society* 37, nos. 1–2 (January–February 1989), pp. 9–19.

P. Damaske. "Head-Related Two-Channel Stereophony with Loudspeaker Reproduction." *Journal of the Acoustical Society of America* 50, no. 4 (1971), pp. 1109–1115.

G. Dutton. "The Assessment of Two-Channel Stereophonic Reproduction Performance in Studio Monitor Rooms, Living Rooms, and Small Theatres." *Journal of the Audio Engineering Society* 10, no. 2 (April 1962), pp. 98–105.

J. Eargle. *Sound Recording.* New York: Van Nostrand Reinhold Company, 1976.

M. Gardner. "Some Single and Multiple Source Localization Effects." *Journal of the Audio Engineering Society* 21, no. 6 (July–August 1973), pp. 430–437.

M. Gardner and R. Gardner. "Problems of Localization in the Median Plane-Effect of Pinnae Cavity Occlusion." *Journal of the Acoustical Society of America* 53 (February 1973), pp. 400–408.

M. Gerzon. "Pictures of Two-Channel Directional Reproduction Systems." Preprint No. 1569 (A4), paper presented at the Audio Engineering Society 65th convention, February 25–28, 1980, London.

M. Gerzon. Letter to the Editor, reply to comments on "Spaciousness and Localization in Listening Rooms and Their Effects on the Recording Technique." *Journal of the Audio Engineering Society* 35, no. 12 (December 1987), pp. 1019–1014.

D. Griesinger. "New Perspectives on Coincident and Semi-Coincident Microphone Arrays." Preprint No. 2464 (H4), paper presented at the Audio Engineering Society 82nd convention March 10–13, 1987, London.

D. Griesinger. "Equalization and Spatial Equalization of Dummy Head Recordings for Loudspeaker Reproduction." *Journal of the Audio Engineering Society* 34, nos. 1–2 (January–February 1989), pp. 20–29.

F. Harvey and M. Schroeder. "Subjective Evaluation of Factors Affecting Two-Channel Stereophony." *Journal of the Audio Engineering Society* 9, no. 1 (January 1961), pp. 19–28.

C. Huggonet and J. Jouhaneau. "Comparative Spatial Transfer Function of Six Different Stereophonic Systems." Preprint 2465 (H5), paper presented at the Audio Engineering Society 82nd convention, March 10–13, 1987, London.

S. Koshigoe and S. Takahashi. "A Consideration on Sound Localization." Preprint No. 1132 (L9), paper presented at the Audio Engineering Society 54th convention, May 4–7, 1976.

S. Lipshitz. Letter to the editor. *Audio* (April 1990), p. 6.

E. R. Madsen. "The Application of Velocity Microphones to Stereophonic Recording." *Journal of the Audio Engineering Society* 5, no. 2 (April 1957), p. 80.

S. Mehrgardt and V. Mellert. "Transformation Characteristics of the External Human Ear." *Journal of the Acoustical Society of America* 61, no. 6, p. 1567.

H. Moller. "Reproduction of Artificial-Head Recordings Through Loudspeakers." *Journal of the Audio Engineering Society* 37, nos. 1–2 (January–February 1989), pp. 30–33.

T. Mori, G. Fujiki, N. Takahashi, and F. Maruyama. "Precision Sound-Image-Localization Technique Utilizing Multi-Track Tape Masters." *Journal of the Audio Engineering Society* 27, nos. 1–2 (January–February 1979), pp. 32–38.

K. Nakabayashi, "A Method of Analyzing the Quadraphonic Sound Field." *Journal of the Audio Engineering Society* 23, no. 3 (April 1975), pp. 187–193.

F. Rumsey. *Stereo Sound for Television.* Boston: Focal Press, 1989.

N. Sakamoto, T. Gotoh, T. Kogure, and M. Shimbo. "On the Advanced Stereophonic Reproducing System 'Ambience Stereo.'" Preprint No. 1361 (G3), paper presented at the Audio Engineering Society 60th convention, May 2–5, 1978, Los Angeles.

N. Sakamoto, T. Gotoh, T. Kogure, and M. Shimbo. "Controlling Sound-Image Localization in Stereophonic Reproduction, Part I." *Journal of the Audio Engineering Society* 29, no. 11 (November 1981), pp. 794–799.

N. Sakamoto, T. Gotoh, T. Kogure, and M. Shimbo. "Controlling Sound-Image Localization in Stereophonic Reproduction, Part II." *Journal of the Audio Engineering Society* 30, no. 10 (October 1982), pp. 719–722.

M. Schroeder and B. Atal. "Computer Simulation of Sound Transmission in Rooms." *IEEE Convention Record,* Part 7 (1963), pp. 150–155.

E. Shaw. "Transformation of Sound Pressure Levels from the Free Field to the Eardrum in the Horizontal Plane." *Journal of the Acoustical Society of America* 56, no. 6 (December 1974), pp. 1848–1861.

G. Theile. "On the Stereophonic Imaging of Natural Spatial Perspective Via Loudspeakers: Theory." In *Perception of Reproduced Sound 1987,* ed. Soren Bech and O. Juhl Pedersen. Munich: Institut fur Rundfunktechnik, 1987.

P. Vanderlyn. British patent 781,186 (August 14, 1957).

P. Vanderlyn. "Auditory Cues in Stereophony." *Wireless World* (September 1979), pp. 55–60.

H. Wallach, E. Newman, and M. Rozenzweig. "The Precedence Effect in Sound Localization." *Journal of the Audio Engineering Society* 21, no. 10 (December 1973), pp. 817–826.

M. Williams. "Unified Theory of Microphone Systems for Stereophonic Sound Recording." Preprint No. 2466 (H-6), paper presented at the Audio Engineering Society 82nd convention, March 10–13, 1987, London.

10

SPECIFIC FREE-FIELD STEREO MICROPHONE TECHNIQUES

Some stereo microphone techniques work better than others. Each method has different effects. A few techniques provide sharper imaging, some create a narrow stage width, some have exaggerated separation, and so on. In this chapter, I compare the characteristics of several specific stereo microphone techniques. All of these use free-field microphones; the next chapters cover stereo techniques using boundary microphones and dummy heads.

Localization Accuracy

One characteristic that varies among different types of arrays is localization accuracy. Localization is accurate if instruments at the sides of the ensemble are reproduced from the left or right speaker, instruments halfway off center are reproduced halfway between the center and one speaker, and so on. In other words, there is little or no distortion of the geometry of the musical ensemble.

For example, suppose your stereo speakers are spaced the same distance apart as you're sitting from them, so that each speaker is ±30° off center. (This is the recommended arrangement for good stereo.) If the orchestral width "seen" by the microphone pair is 90°, we want sources that are 45° to one side of center to be reproduced out of only one speaker. Sources 22.5° off center should be reproduced halfway between the center of the speaker pair and one speaker (15° off center).

Figure 10–1 illustrates this. In Figure 10–1(a), the letters A through E represent live sound source positions relative to the microphone pair. In Figure 10–1(b), the corresponding images of these sources are accurately localized between the speaker pair.

Spacing or angling the microphones more than is necessary to achieve a full stereo spread produces an "exaggerated separation" effect: Instruments near the center are reproduced to the extreme left or right, rather than slightly off center. Instruments exactly in the center are still reproduced between the speakers (see Figure 10–1(c)). Conversely, too little angling or spacing gives a poor stereo spread, a "narrow stage-width" effect (see Figure 10–1(d)).

A listening test was performed to determine the localization accuracy of various stereo microphone techniques, for a 90° orchestral width (Bartlett, 1979). Recordings were made of a speech source at 0°, 22.5°, and 45° relative to the microphone pair (as in Figure10–2(a)).Tests were made in an anechoic chamber and a reverberant gymnasium. Listeners were asked to note the reproduced sound-image locations for several techniques. The image locations of the anechoic and reverberant recording rooms were averaged, with results shown in Figure 10–2(b).

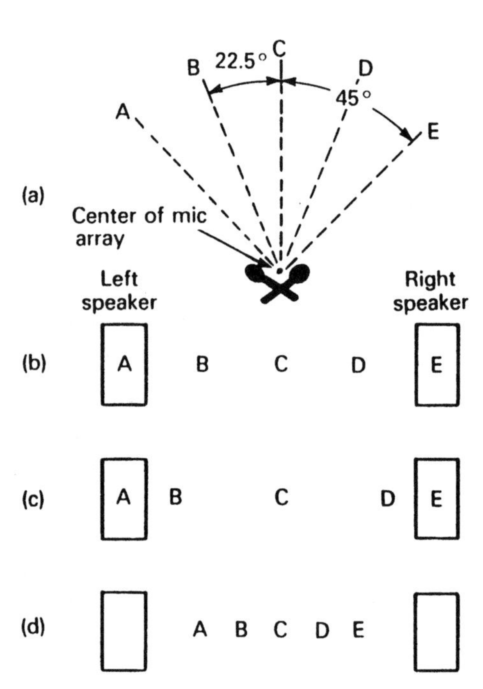

Figure 10–1
Stereo localization effects for a 90° (±45°) orchestral width: (a) letters A through E represent live sound-source positions (top view), (b) accurately localized images between speakers (listener's perception), (c) exaggerated separation effect, (d) narrow stage width effect.

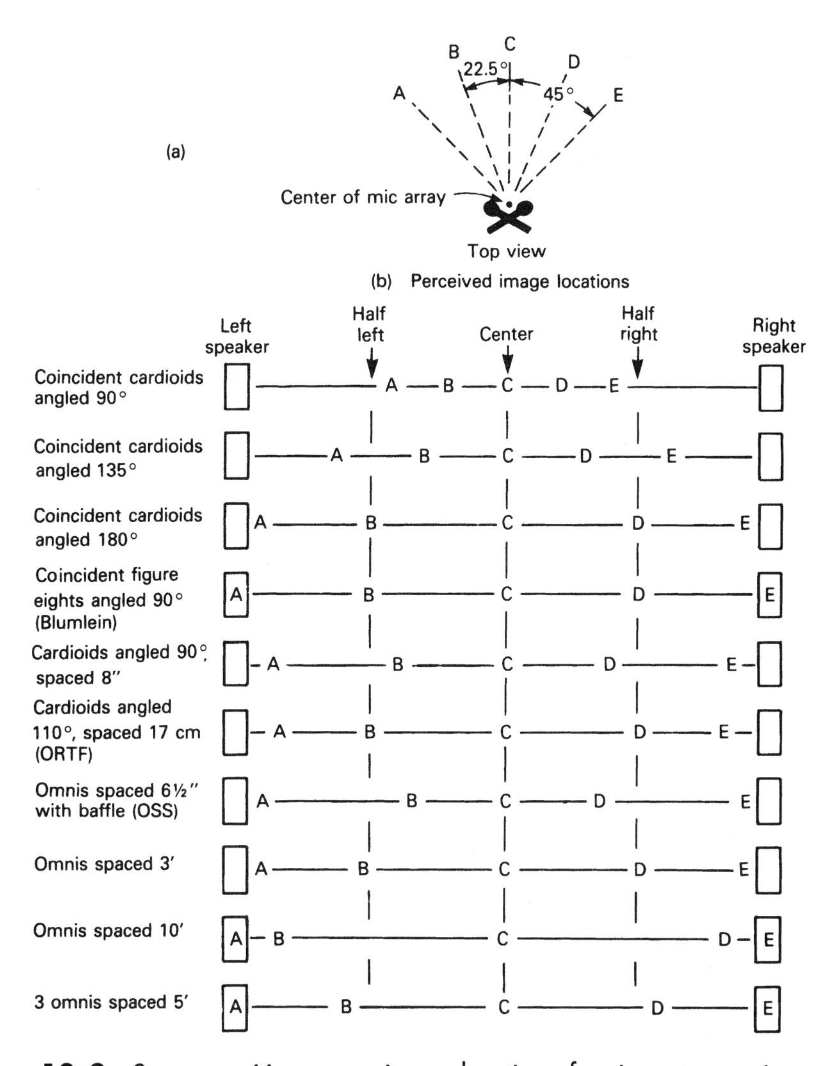

Figure 10–2 Source position versus image location of various stereo mic arrays: (a) letters A through E are live speech-source positions relative to the mic array, (b) stereo image localization of various stereo mic arrays (listener's perception). Images A through E correspond to live speech sources A through E in (a).

Since results may vary under different listening conditions, this information is meant to be indicative rather than definitive. Different listeners hear stereo effects differently, so your perceptions may not agree exactly with those shown. Still, Figure 10–2 lets you compare one technique to another.

The 90° orchestral width used is arbitrary. The actual width of the orchestra varies with the size of the ensemble and the mic-to-source distance. If the orchestral width is more than 90°, the stereo spread of all these techniques is wider than shown in Figure 10–2(b).

The closer to the ensemble a microphone array is placed, the greater is the orchestral width as seen by the microphone pair and, so, the wider is the stereo spread (up to the limit of the speaker spacing).

Examples of Coincident-Pair Techniques

In general, coincident cardioids tend to give a narrow stereo spread and lack a sense of air or spaciousness. Imaging at high frequencies is not optimum because there is no time difference between channels, which, according to Cooper, is essential. Also, when microphones are angled apart, they receive much of the sound off axis. Many microphones have off-axis coloration (a different frequency response on and off axis).

Coincident techniques are mono-compatible: The frequency response is the same in mono and stereo. That is because there are no phase or time differences between channels to cause phase cancellations if the two channels are mixed to mono.

Coincident Cardioids Angled 180° Apart

According to Figure 10–2(b), it seems reasonable to angle two coincident cardioid microphones 180° apart to achieve maximum stereo spread (as shown in Figure 10–3). However, sounds arriving from straight ahead approach each microphone 90° off axis. The 90° off-axis frequency response of some microphones is weak in high frequencies, giving a dull sound to instruments in the center of the orchestra. In addition, it has been the experience of another experimenter, Michael Gerzon (1976, p. 36), that 180° angling places the reproduced reverberation to the extreme left and right.

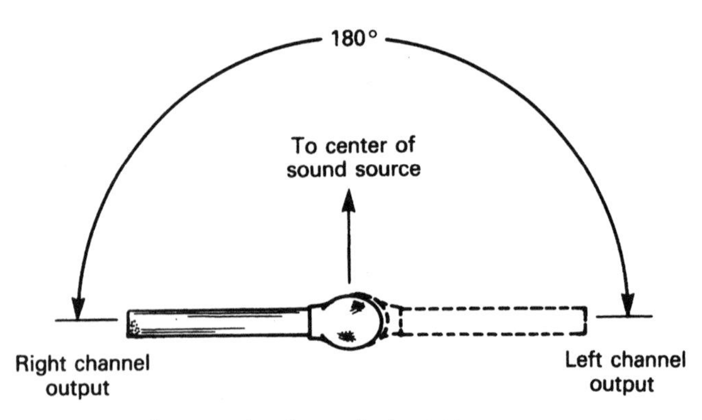

Figure 10–3 Coincident cardioids angled 180° apart.

Coincident Cardioids Angled 120–135° Apart

A 120–135° angle between microphones might be a better compromise. Gerzon has reported that the 120° angle gives a uniform spread of reverberation between speakers, while the 135° angle (Figure 10–4) provides a slightly wider stereo spread. These angles are useful where maximum stereo spread of the source is not desired. For a wider stereo spread, you can use a near-coincident or spaced pair. However, the 135° angle just described can provide a full stereo spread if the orchestral width or source angle is 150°.

Coincident Cardioids Angled 90° Apart

Angling cardioids at 90° (Figure 10–5) reproduces most of the reverberation in the center. It gives a narrow stage width, unless the ensemble surrounds the microphone pair in a semicircle (180° source angle).

Blumlein or Stereosonic Technique (Coincident Bidirectionals Angled 90° Apart)

This technique is diagrammed in Figure 10–6. As shown in Figure 10–2(b), it provides accurate localization. According to Gerzon (1976) and the listening tests, it also provides sharp imaging, a fine sense of depth, and the most uniform possible spread of reverberation across the reproduced stereo stage. It has the sharpest perceived image focus of any system, other than spatially equalized systems (Huggonet and Jouhaneau, 1987, Figure 8, p. 11).

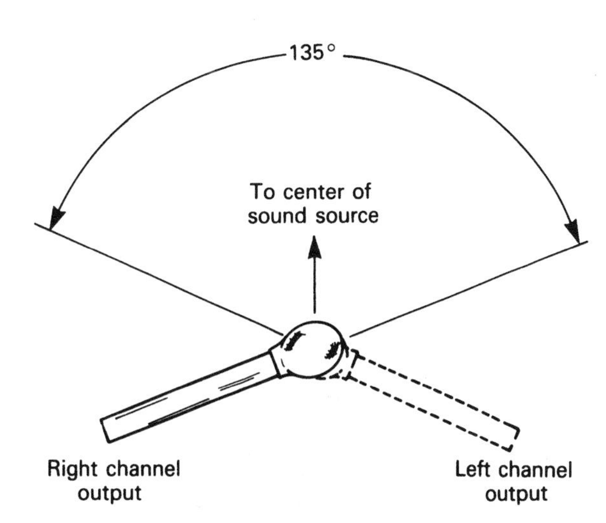

Figure 10–4
Coincident cardioids angled 135° apart.

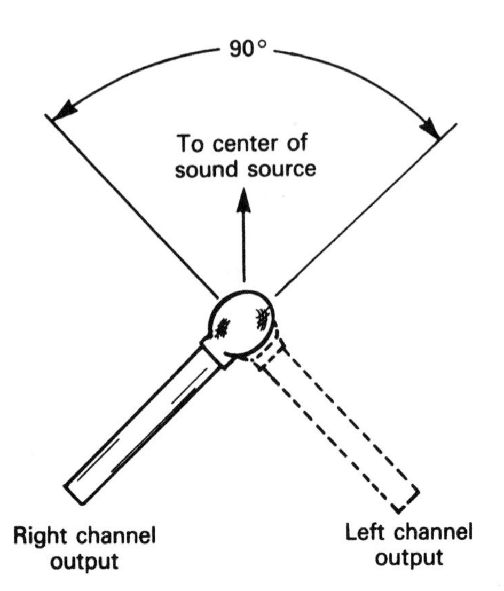

Figure 10–5
Coincident cardioids angled 90° apart.

Note that each bidirectional pattern has a rear lobe in opposite polarity to the front lobe. If a sound source is more than 45° off center (say, off to the left side), it is picked up by the front-left lobe and the back-right lobe. These are opposite in polarity. This creates antiphase information between channels, which produces vague localization. For this reason, the microphones should aim at the extreme-left and -right ends of the performing ensemble. This prevents sound sources from being outside the 45° limit. However, this limitation fixes the mic-to-source distance. You can't adjust this distance to vary the sense of perspective unless you also change the angle between microphones or the size of the musical ensemble.

Another drawback is that the microphones pick up a large amount of reverberation. If you place the microphone pair closer to the ensemble to increase the direct/reverb ratio, the stereo spread becomes excessive and

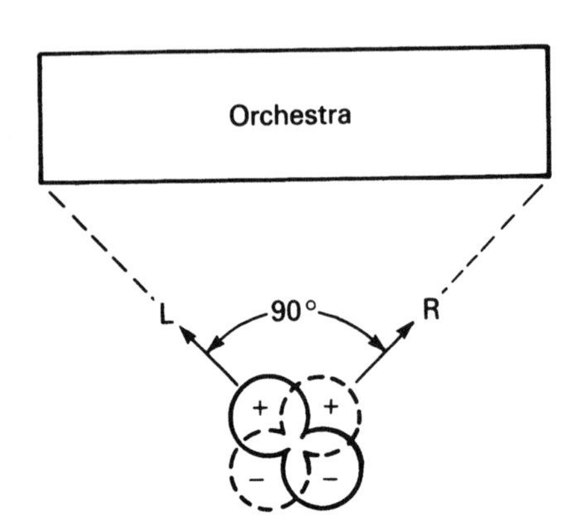

Figure 10–6
The Blumlein or stereosonic technique (coincident bidirectionals crossed at 90°).

instruments in the center of the ensemble are emphasized. In addition, instruments at either end of the ensemble are reproduced with opposite polarity signals from both channels, so they are not localized.

The Blumlein technique works best in a wide room with minimal sidewall reflections, where strong signals are not presented to the sides of the stereo pair (Streicher and Dooley, 1985a).

Hypercardioids Angled 110° Apart

Shown in Figure 10–7, this method gives accurate localization. Listening tests also reveal sharp imaging and very good spaciousness. This array has the widest in-phase region of any array that has a spaciousness of 1 (Griesinger, 1987). The tight pattern of the hypercardioid allows a more distant placement than with crossed cardioids. As for drawbacks, hypercardioid microphones tend to have a bass roll-off; but this can be corrected with equalization (bass boost).

XY Shotgun Microphones

Henning Gerlach of Sennheiser Electronics suggests that two shotgun microphones can be crossed in an XY configuration with their diaphragms coincident (1989). To avoid exaggerated separation, the angle between microphones must be small. Image positioning with handheld shotguns is unstable, so this method is recommended only for stationary sound sources and stationary microphones.

Another coincident technique is the mid-side (MS) technique, which will be covered in detail later in this chapter.

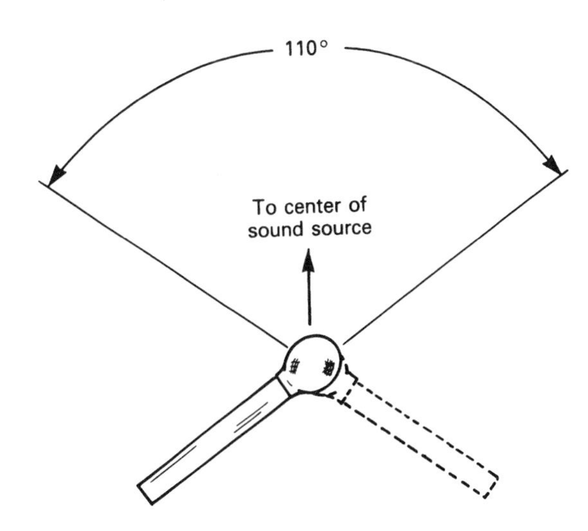

Figure 10–7
Hypercardioids angled 110° apart.

Examples of Near-Coincident-Pair Techniques

Near-coincident-pair techniques give wider stereo spread than coincident-pair techniques that have the same angle between mics. In other words, if you start with a coincident-pair technique and space the mics a few inches apart, the stereo spread will increase. Near-coincident-pair methods also have more spaciousness and depth. This is due to the random phase relationships (low correlation) between channels at high frequencies.

These methods are not mono-compatible: If both channels are combined to mono, there are dips in the frequency response caused by phase cancellations. And, since the microphones are angled apart, the sound source might be reproduced with off-axis coloration.

The ORTF System: Cardioids Angled 110° Apart and Spaced 17 cm (6.7 inches) Horizontally

The listening tests summarized in Figure 10–2(b) reveal that the 110° angled, 17 cm spaced array (the ORTF, French Broadcasting Organization, system) and the 90° angled, 8 inch spaced array tend to provide accurate localization. These two methods are shown in Figures 10–8 and 10–9. According to a listening test conducted by Carl Ceoen (1972), the ORTF system was preferred over several other stereo-miking techniques. It provided the best overall compromise of localization accuracy, image sharpness, an even balance across the stage, and ambient warmth.

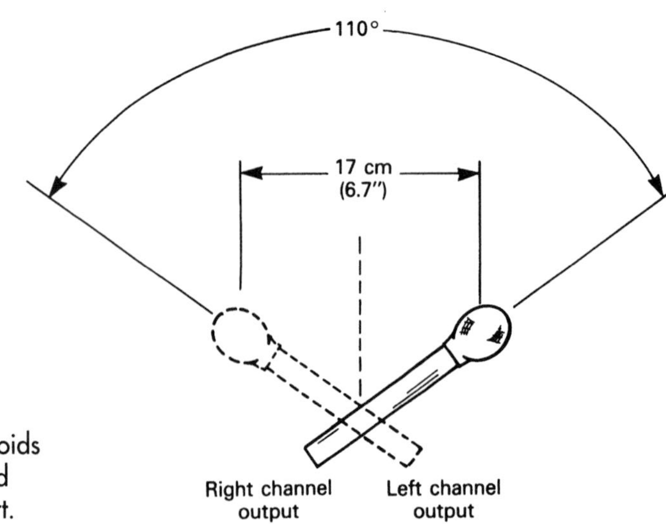

Figure 10–8
The ORTF system: Cardioids angled 110° and spaced 17 cm (6.7 inches) apart.

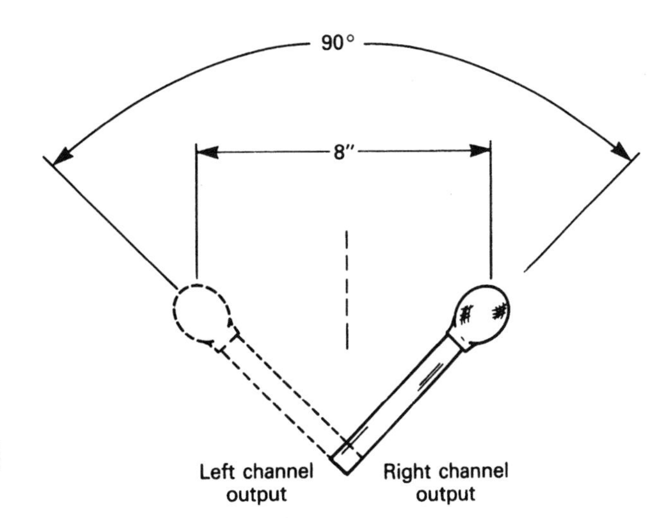

Figure 10–9
Cardioids angled 90° and spaced 8 inches apart.

The origin of the ORTF system was described by R. Condamines (1978). The 17 cm spacing was chosen because it provided the best image stability with head motion, assuming a speaker angle of ±30°. The 110° angle was chosen because it provided the best image precision and placement when used with a 17 cm spacing. Condamines reported that, if the mic angle is less than 110°, the sound stage usually does not spread all the way between speakers; if the angle is greater than 110°, the center image becomes weak (a hole-in-the-middle effect).

The ORTF image position varies with frequency, according to calculation (Bernfeld and Smith, 1978) and perception (Huggonet and Jouhaneau, 1987, p. 14, Fig. 11).

The NOS System: Cardioids Angled 90° Apart and Spaced 30 cm (12 inches) Horizontally

Shown in Figure 10–10, this system was proposed by the Dutch Broadcasting Foundation. Since the spacing of the NOS system exceeds the 90° angled, 8 inch spaced array in the listening test, we could expect it to have a slightly wider stereo spread for halfway left and halfway right instruments.

The OSS (Optimal Stereo Signal) or Jecklin Disk

The Jecklin disk uses two omnidirectional microphones spaced 16.5 cm (6.5 inches) apart and separated by a disk with a diameter of 28 cm ($11\frac{7}{8}$ inches) (Jecklin, 1981). The disk is hard and is covered with absorbent

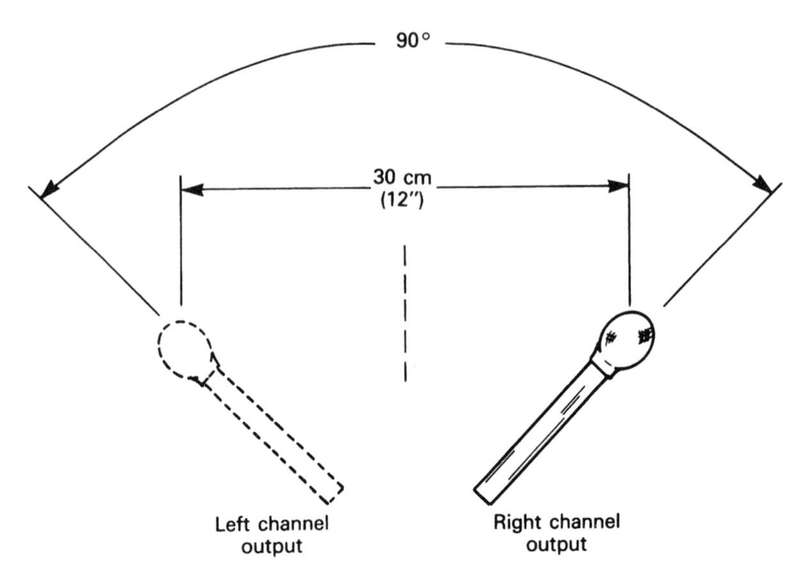

Figure 10–10 The NOS system: cardioids angled 90° and spaced 30 cm (12 inches) apart.

material to reduce reflections (Figure 10–11). The OSS system could be called *quasi-binaural,* in that the human binaural hearing system also uses two omni "microphones" separated by a baffle (the head).

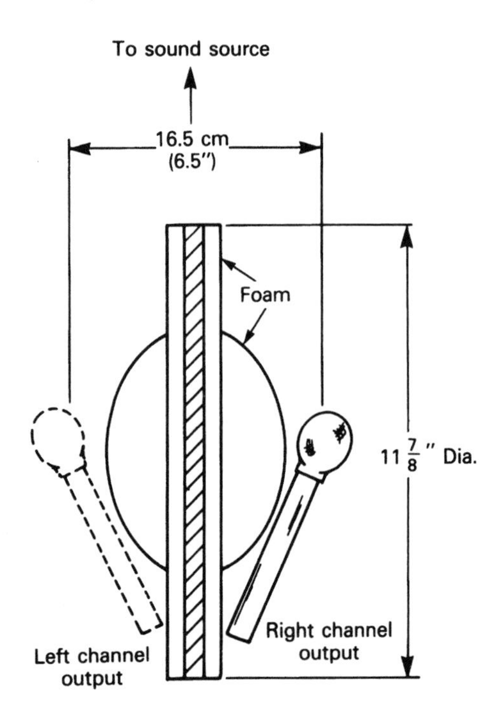

Figure 10–11
The OSS system or Jecklin disk. Omnis are spaced 16.5 cm (6.5 inches) apart and separated by a foam-covered disk of 28 cm ($11\frac{7}{8}$ inches) diameter.

Below 200 Hz, both microphones receive the same amplitude, and the array acts like closely spaced omnis. As frequency increases, the disk becomes more of a sound barrier, which makes the array increasingly directional. At high frequencies, the array acts like near-coincident-pair subcardioids angled 180° apart.

Since both channels receive the same signal level at low frequencies, stereo localization at low frequencies can be due only to the capsule spacing, which causes direction-dependent delays. But, according to Griesinger (1987), delay panning does not create localizable images below 500 Hz. If that is true, the OSS system localizes only above 200 Hz.

According to the inventor, "the stereo image is nearly spectacular, and the sound is rich, full, and clear." It "seems to be superior to all other recording methods." The full sound is probably due to the use of omnidirectional condenser microphones, which have an extended low-frequency response.

Listening tests (Figure 10–2(b)) show that the OSS stereo spread for a 90° orchestral width is somewhat narrow. But, since the system uses omni microphones, it usually is placed close to the ensemble, where the angular width of the ensemble is wide. This results in a wider stereo spread. The unit is not mono-compatible.

Examples of Spaced-Pair Techniques

In general, listeners commented that the spaced-pair methods give relatively vague, hard-to-localize images for off-center sources. These methods are useful when you want diffuse images for special effect. Spaced arrays have a pleasing sense of spaciousness. This is produced artificially by the random phase relationships between channels and opposite-polarity signals at various frequencies (Lipshitz, 1986).

Spaced-pair techniques are not mono-compatible: Peaks and dips in the frequency response of the direct sound occur when both channels are combined to mono. This effect may or may not be audible, because reverberation approaches the microphones from all angles and each angle of sound incidence relates to a different pattern of phase cancellations. The reverberation randomizes the frequencies of these cancellations, so that the effect is less audible.

The spaced array has extreme phase differences between channels, which can make record cutting difficult due to excess vertical modulation of the record groove by out-of-phase components. Also, the array is relatively big and unwieldy.

An advantage of the spaced-pair technique is that it allows the use of omnidirectional condenser microphones, which have a more extended low-frequency response than directional microphones. That is, the tone quality is warmer and fuller in the bass. Of course, you can equalize directional microphones to have a flat bass response at a distance.

Another advantage is that the listening area for good stereo is wider than with coincident-pair techniques. The spaced-pair delay cues counteract the amplitude imbalance that occurs when the listener sits off center.

Many instruments, such as the flute, have nulls in their radiation pattern that vary with the note played. Thus, one mic of a spaced pair might pick up a note at a low level, while the other mic would pick it up at a high level, so the image would wander with the note played. However, one mic will pick up notes that the other mic misses. Our ears have the same ability due to their spacing. Thus, the spaced-pair method offers the potential for better fidelity (no missed notes) at the expense of wandering images (Lemon, 1989).

You can use cardioids or other unidirectional patterns in a spaced array to reduce the pickup of hall reverberation. These patterns, however, tend to have less bass than omnis. Spaced bidirectionals have very little off-axis coloration.

Omnis Spaced 3 Feet Apart

Shown in Figure 10–12, this method gives fairly accurate localization (Figure 10–2(b)) but with poorly focused imaging of off-center sources. A

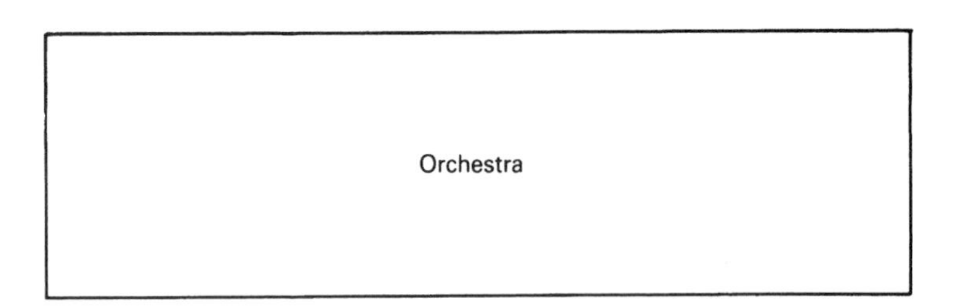

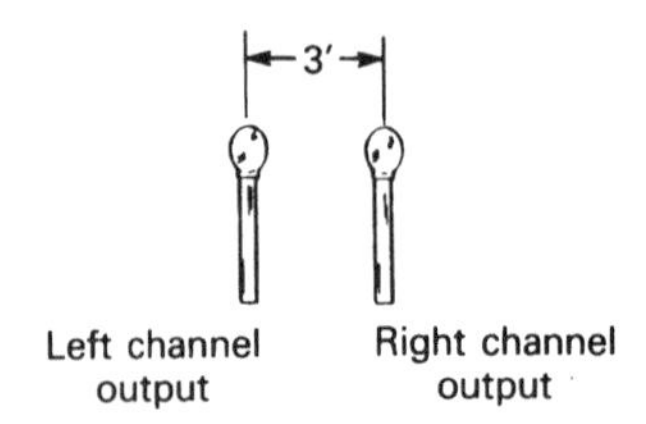

Figure 10–12 Omnis spaced 3 feet apart.

2 foot spacing would give more accurate localization. Since omnis must be placed relatively close to a performing ensemble for an acceptable direct/reverb ratio, this array is likely to overemphasize the center instruments. That is, the microphone pair is most sensitive to instruments in the center of the orchestra, with reduced pickup of the sides.

Telarc often uses a 2 foot spaced pair, angled 90° to each other, about 10 feet high, plus a pair of flanking omnis spaced 10–15 feet each side of center. The flanking mics are 2–3 dB below the center pair. The center mics are panned partly left and right; the flanks are panned hard left and right.

Omnis Spaced 10 Feet Apart

Shown in Figure 10–13, this spacing provides a more even coverage of the orchestra (a better balance). However, spacings greater than 3 feet give an exaggerated separation effect, in which instruments slightly off center are reproduced full-left or full-right (Figure 10–2(b)). This dispels the myth that spaced microphones should be as far apart as the playback loudspeakers. Instruments directly in the center of the ensemble are still reproduced exactly between the speakers.

Three Omnis Spaced 5 Feet Apart (10 Feet End to End)

With this method (Figure 10–14), a third microphone is placed between the other two, mixed in at an approximately equal level and split to both channels. This reduces stereo separation while maintaining full coverage of the orchestra (see Figure 10–2). The three-spaced-omnis technique often is used by Telarc Records. Image focus and mono-compatibility are fair to good.

Decca Tree

Developed in 1954 by the Decca Record Company, the Decca Tree is an array of three spaced omnidirectional mics (Figure 10–15; Gayford, 1994; http://www.josephson.com/deccatree). Mic spacing depends on the desired amount of width and spaciousness. The center mic is placed slightly forward of the outer pair. Because the center mic's signal precedes that of the outer pair, the center mic helps to "solidify" the center image.

As for placement, the triangle of mics are mounted about 10–12 feet above the stage, just behind the conductor. The outer pair are angled outward to point at the edges of the stage, so that the edges are picked up with the best high-frequency response. Central sounds are on axis to the center

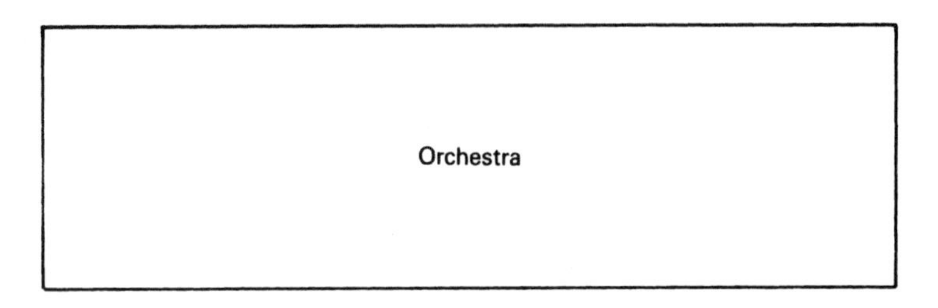

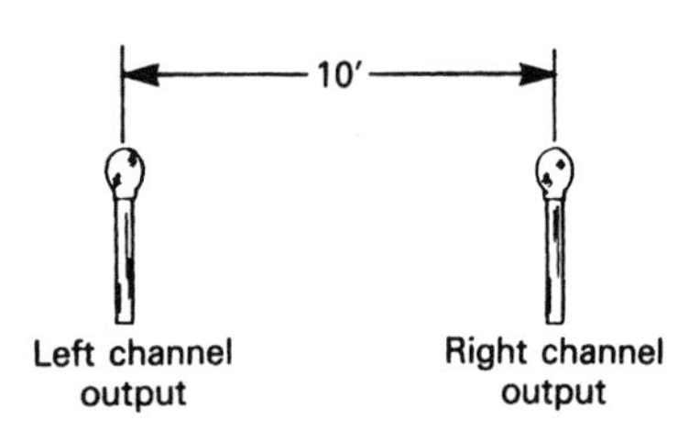

Figure 10–13 Omnis spaced 10 feet apart.

microphone. The center mic may exacerbate the comb-filtering effects that sometimes occur with spaced pairs.

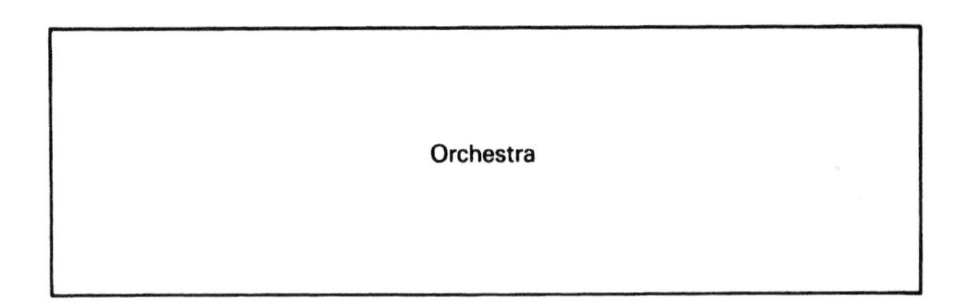

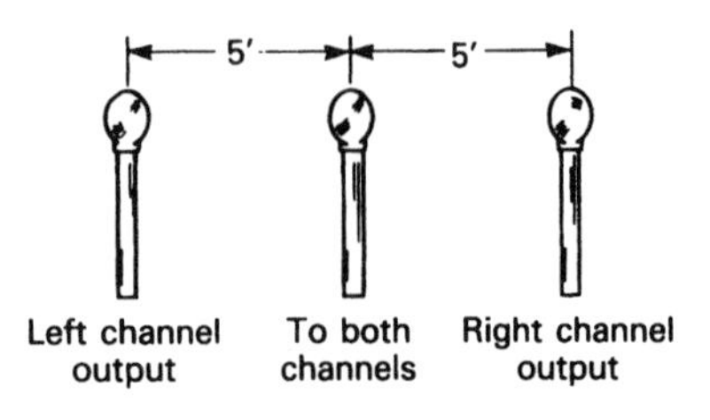

Figure 10–14 Three omnis spaced 5 feet apart.

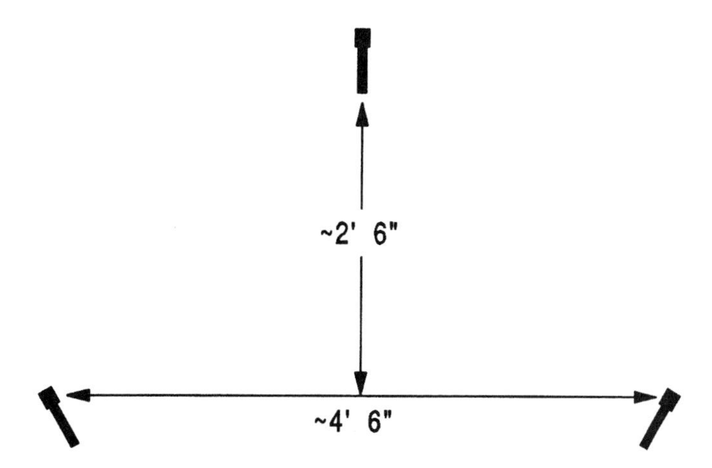

Figure 10–15 Decca tree stereo microphone technique.

Sometimes, an additional pair of flanking mics are used near the edges of the orchestra or about one third of the way in. These flanking mics face diagonally across the orchestra and help to add width and spaciousness. All mics are mixed at an equal level. The center mic is panned to center, both left mics are panned hard left, and both right mics are panned hard right.

Mic spacing varies with the venue and the ensemble size. The center mic or the outriggers might be omitted in some cases.

Other Coincident-Pair Techniques

Let's return to coincident-pair methods and go over some specific techniques in detail.

MS (Mid-Side)

This method uses a middle (mid) microphone capsule aiming straight ahead toward the center of the performing ensemble plus a side-aiming (side) bidirectional microphone capsule. These capsules are coincident and at right angles to each other (as shown in Figure 10–16). The middle capsule is most commonly cardioid, but it can be any pattern.

The outputs of both capsules are summed to produce the left-channel signal and are differenced to produce the right-channel signal. In effect, this creates two virtual polar patterns angled apart:

$$M + S = L$$
$$M - S = R$$

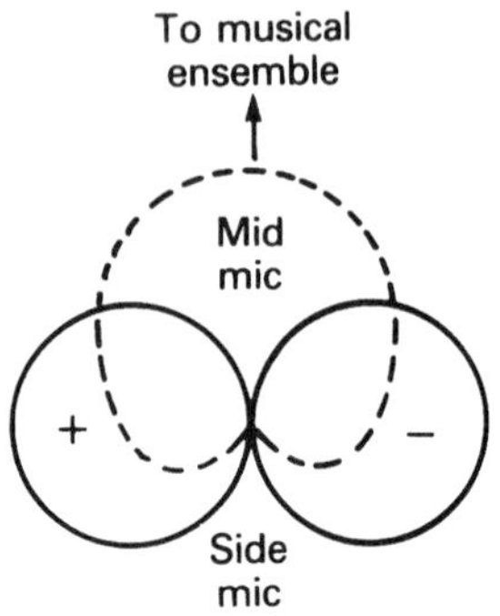

Figure 10–16
Mid-side (MS) stereo microphone technique.

For example, suppose that the mid capsule is omnidirectional and the side capsule is bidirectional. Also suppose that the sensitivity of both capsules is set equal. When you add these two patterns together, you get a cardioid aiming 90° to the left. When you subtract these patterns (add them in opposite polarity) you get a cardioid aiming 90° to the right. Thus, a mid-side microphone with an omni mid capsule is equivalent to two coincident cardioids angled 180° apart. A mid-side microphone with a bidirectional mid capsule is equivalent to two figure eights crossed at 90° (the Blumlein technique).

Some stereo microphones have switchable polar patterns. Changing the mid-capsule pattern changes the pattern and angling of the virtual polar patterns. The more directional the mid mic is, the more directional are the sum-and-difference (virtual) polar patterns (Figure 10–17). Consequently, you can change the apparent distance from the sound source by changing the mid pattern.

MS Matrix Box

The M and S outputs of the microphone are connected to an MS matrix box or decoder. This decoder uses either a tapped transformer or an active circuit (such as that shown in Figure 10–18) to sum and subtract the M and S signals. The output of the box is a left-channel signal and a right-channel signal. A schematic for an active matrix was reported by Pizzi (1984).

A rotating knob in the box controls the ratio of the mid signal to the side signal. By varying the ratio of mid to side signals, you change the polar pattern and angling of the left and right virtual mic capsules. In turn, this varies the stereo spread and the ratio of direct-to-reverberant sound. As you turn up the side signal, the stereo spread widens and the ambience

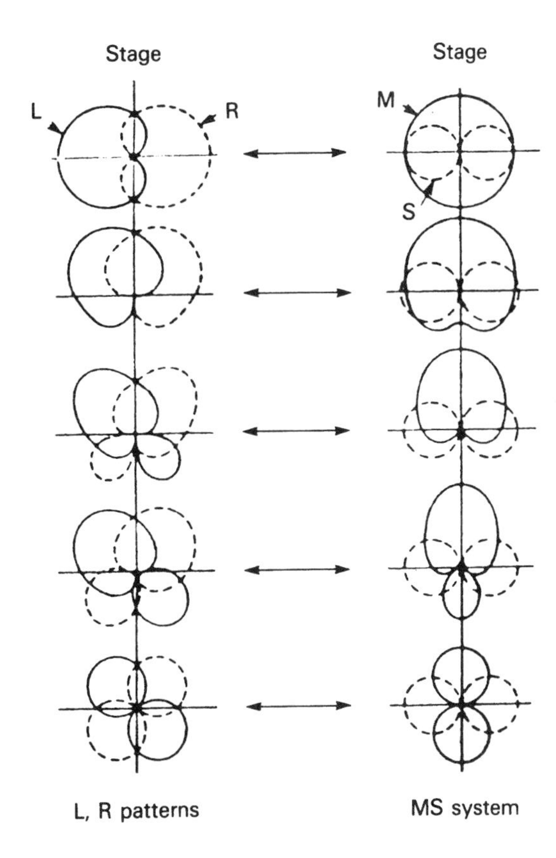

Figure 10–17
Equivalent directional patterns for MS system, with mid pattern varied. (From a letter by Les Stuck to *db* magazine, March 1981.)

increases, as shown in Figure 10–19. The optimum starting M/S ratio is near 1:1.

By using two matrix boxes in series, you can vary the spread of any standard left/right stereo mic technique. An alternative to the matrix

Figure 10–18
The Audio Engineering Associates MS38 active matrix decoder for an MS stereo. (Courtesy of Audio Engineering Associates.)

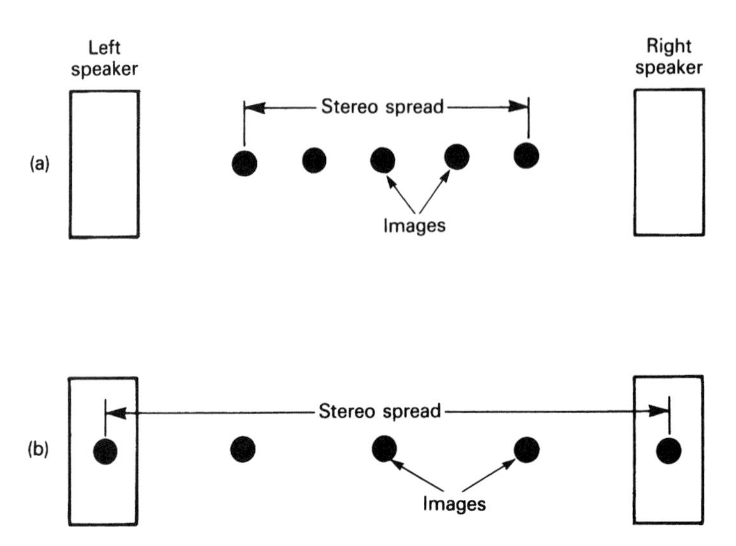

Figure 10–19 Effects of varying M/S ratio on stereo spread: (a) high M/S ratio gives a narrow spread, (b) low M/S ratio gives a wide spread.

decoder is a mixer. As shown in Figure 10–20, you pan the M signal to the center and split the S signal to two inputs with the polarity inverted in one input. To do this, reverse the connections to pins 2 and 3 in one mic-cable connector. Pan the out-of-phase (opposite polarity) S signals hard left and right. You can vary the mid/side ratio with the microphone fader pairs.

MS Advantages

A major advantage of the MS system is that you can control the stereo spread from a remote location. This feature is especially useful for live concerts, where you can't change the microphone array during the concert. Since the stereo spread is adjustable, the MS system can be made to have accurate localization.

If you record the M and S signals directly to a two-track recorder during the concert, you can play them back through a matrix decoder after the concert and adjust the stereo spread then. In postproduction, you can vary the spread from very narrow (mono) to very wide. While recording the concert, you monitor the outputs of the matrix decoder but do not record them.

The MS method has another advantage: It is fully mono-compatible. If you sum the left and right channels to mono, you get just the output of the forward-facing mid capsule. This is shown in the following equations:

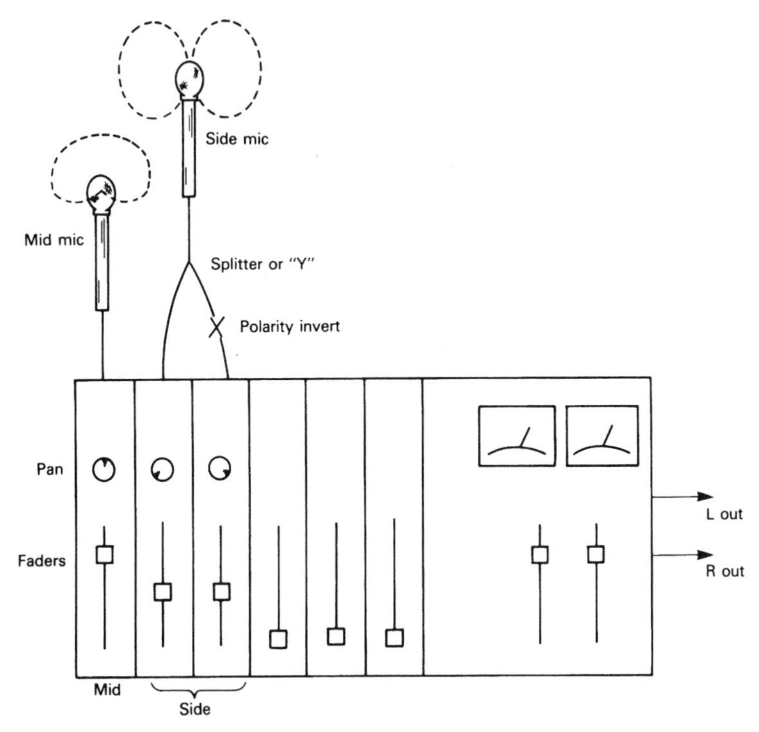

Figure 10–20 A method of using a stereo mixer as an MS matrix decoder.

$$\text{Left} = (M + S)$$
$$\text{Right} = (M - S)$$
$$\text{Left} + \text{Right} = (M + S) + (M - S) = 2M$$

With XY or near-coincident techniques, the center image is formed by adding the output of two angled directional capsules. If they are not perfectly matched in frequency response and phase response, the fusion of the center image can be degraded. But the MS system has very sharp center imaging because the center image is the output of the single mid capsule.

MS Disadvantages

The MS system has been criticized for a lack of warmth, intimacy, and spaciousness (Ceoen, 1972; Griesinger, 1987). However, Griesinger states that MS recording can be made more spacious by giving the low frequencies a shelving boost of 4 dB (+2 dB at 600 Hz) in the L – R or side signal, with a complementary shelving cut in the L + R or mid signal.

There are other disadvantages to the MS technique. It requires a matrix decoder, which is extra hardware to take on location. If you record the MS signals on analog tape for decoding during playback, an extra tape generation is required to record the left/right decoded signals. This analog generation degrades the sound quality. A final disadvantage is that the stereo spread and direct-to-reverb ratio are interdependent: You can't change one without changing the other.

When the signals from an MS stereo microphone are mixed to mono, the resulting signal is only from the front-facing mid capsule. If this capsule's pattern is cardioid, sound sources to the far left or right will be attenuated. Thus, the balance might be different in stereo and mono. If this is a problem, use an XY coincident pair rather than MS.

Double MS Technique

Skip Pizzi recommends a double MS technique, which uses a close MS microphone mixed with a distant MS microphone. One MS microphone is close to the performing ensemble for clarity and sharp imaging, and the other is 50–75 feet out in the hall for ambience and depth. The distant mic could be replaced by an XY pair for lower cost (Pizzi, 1984).

For a comprehensive discussion of the MS system, see Streicher and Dooley (1985b).

MS with a Mid Shotgun Mic

Henning Gerlach (1989) of Sennheiser Electronics has suggested that the MS method could be used with a shotgun microphone as the mid element. He notes drawbacks to this method. The stereo spread decreases with frequency. Also, the technique is satisfactory only if the sound source is fairly close to the axis of the shotgun microphone. When you follow the main source with the shotgun, the side capsule should remain stationary or else the stereo image will shift.

Gerlach offers a way around these limitations. Use a standard MS configuration, and mix a boom-mounted, movable shotgun mic with the M signal. Use the shotgun to pick up the main source.

SoundField Microphone

This British microphone (shown in Figure 10–21) is an elaboration on the MS system. It uses four closely spaced cardioid mic capsules arranged in a tetrahedron and aiming outward. Their output is phase shifted to make the capsules seem perfectly coincident.

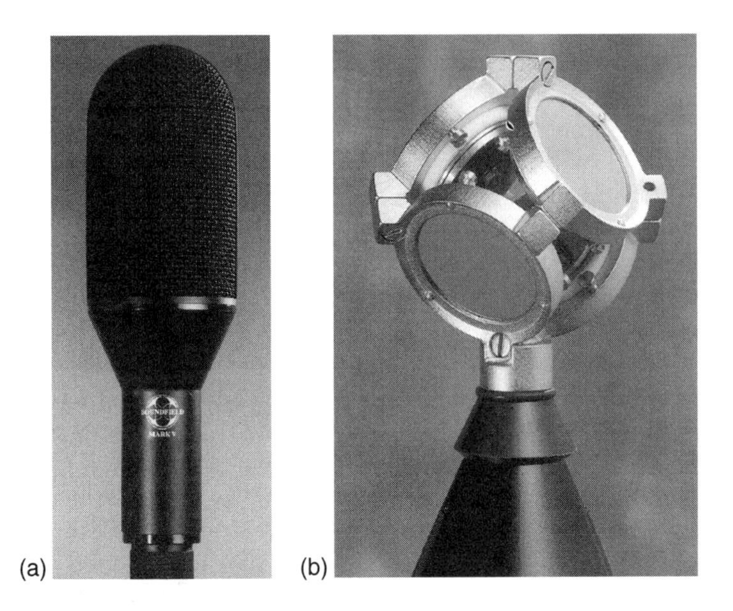

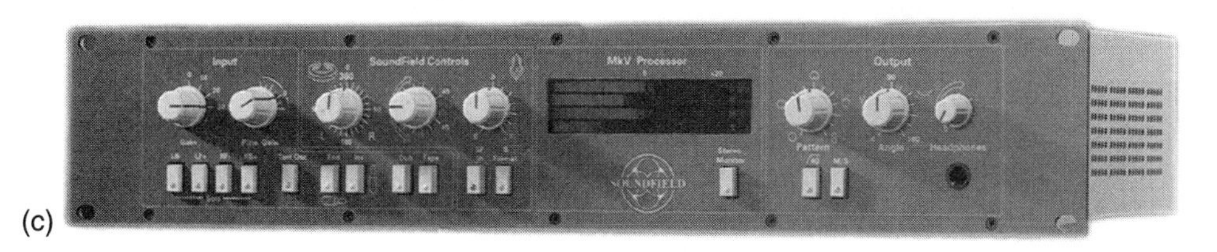

Figure 10–21 SoundField Mk V Microphone: (a) external view; (b) internal view, showing capsules; (c) the Mk V control unit. (Courtesy of SoundField.)

The capsule output is called the *A-format signals.* They are electronically matrixed to produce

- An omnidirectional component (the sound pressure).
- A vertical pressure-gradient component.
- A left-right pressure-gradient component.
- A fore-aft pressure-gradient component.

These B-format signals can be further processed into stereo, quadraphonic, or ambisonic signals. Ambisonic signals include height, fore-aft, and left-right information. With a remote-control box, the user can adjust polar patterns, azimuth (horizontal rotation), elevation (vertical tilt), dominance (apparent distance), and angle (stereo spread) (Streicher and Dooley, 1985; Farrar, 1979).

As for drawbacks, the microphone system costs over $5000 and requires a complex matrix circuit. But it is the world's premier microphone for spatial recording. Several models are available.

The SoundField 5.1 Microphone System is a single, multiple-capsule microphone (SoundField ST250 or Mk V) and SoundField Surround Decoder for recording in surround sound. The decoder translates the mic's B-format signals (X, Y, Z, and W) into L, C, R, LR, RR, and mono subwoofer outputs.

Coincident Systems with Spatial Equalization (Shuffler Circuit)

Coincident-pair systems have been criticized for a lack of spaciousness. However, as discovered by Blumlein (1958), Vanderlyn (1954), and Griesinger (1986, 1987), the focus and spaciousness can be improved by a shuffler circuit (spatial equalization). This circuit decreases stereo separation at high frequencies or increases separation at low frequencies to align the image locations at low and high frequencies. To increase low-frequency separation, the circuit applies a shelving boost to low frequencies in the left-minus-right (difference) signal, and applies a complementary cut to the left-plus-right (sum) signal.

Griesinger reports that spatially equalized coincident or near-coincident arrays have very sharp imaging, and sound as spacious as a spaced array. As stated earlier, MS recordings can be made more spacious by boosting the bass 4 dB (+2 dB at 600 Hz) in the L – R or side signal and cutting the sum signal by the same amount.

Other Near-Coincident Pair Techniques

Let's look more closely at some unusual near-coincident miking methods.

Stereo 180 System

Another near-coincident method is the Stereo 180 System developed by Lynn T. Olson (1979), shown in Figure 10–22. It uses two hypercardioid pattern microphones, angled 135° apart and spaced 4.6 cm (1.8 inches) apart horizontally. The hypercardioid patterns have opposite-polarity rear lobes, which create the illusion that the reproduced reverberation is coming from the sides of the listening room as well as between the speakers. The localization accuracy and image focus of the array are reported to be very good.

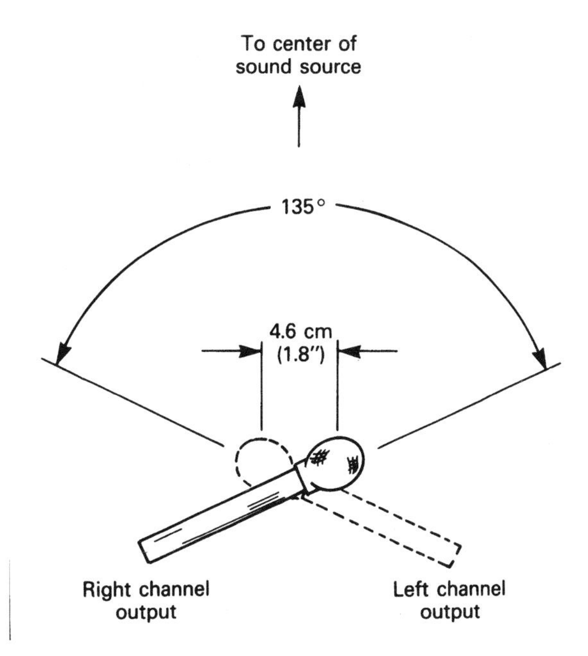

Figure 10–22
Stereo 180 system: hypercardioids angled 135° and spaced 4.6 cm (1.8 inches) apart.

Faulkner Phased-Array System

Invented by Tony Faulkner (1982), this method uses two bidirectional (figure-eight) microphones aiming straight ahead with axes parallel and spaced 20 cm (7.87 inches) apart (Figure 10–23). The plane of maximum path difference coincides with the null in the directional polar pattern of the microphones. Since the microphones are aimed forward rather than angled apart, you can place them farther from the ensemble for a better balance. This distant placement also lets you place the microphones at ear

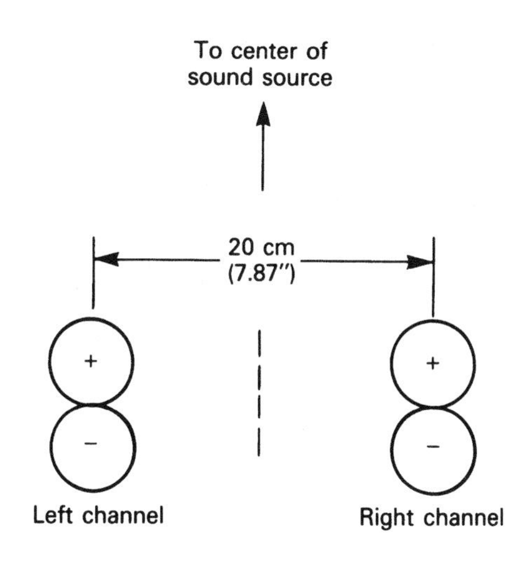

Figure 10–23
Faulkner phased-array system: two figure eights spaced 20 cm (7.87 inches) apart.

height rather than raised. Faulkner says that the array is not mono-compatible in theory but has presented no problems in practice.

Sometimes Faulkner adds a pair of omnidirectional microphones 2–3 feet apart, flanking the figure eights. These omnis add ambient spaciousness.

Near-Coincident Systems with Spatial Equalization

Spatial equalization (described earlier) also improves the image focus of near-coincident methods.

Near-Coincident/Spaced-Pair Hybrid

John Eargle, director of Recording at Delos International Inc., prefers to use a combination of near-coincident and spaced-pair methods (*Symphonic Sound Stage* CD). A quasi-ORTF pair is placed about 4 feet behind the conductor, 9–12 feet high. This pair is flanked by two omnis 12–16 feet apart, typically 6 dB below the main pair. The ORTF pair provides sharp imaging and depth, while the spaced pair adds width to the strings and time cues from the hall. Since the spaced pair uses omnidirectional mics, low-frequency reproduction is excellent.

A second stereo pair is placed up to 30 feet behind the main pair to capture hall reverb. The woodwinds often are picked up with an overhead pair, and accent mics are added if necessary for soloists, harp, celeste, and other instruments.

Comparisons of Various Techniques

Many studies have been done comparing standard stereo miking techniques. The results of some of these are presented here. They do not all agree.

Michael Williams, "Unified Theory of Microphone Systems for Stereophonic Sound Recording" (1987)

Michael Williams calculated the recording angle and standard deviation of several fixed systems. *Recording angle* means the angle subtended by the sound source required for a speaker-to-speaker stereo spread. The angular width of the performing ensemble (as seen by the microphone array) causes a full stereo spread.

The standard deviation represents geometric distortion of the sound stage. The bigger the standard deviation in degrees, the wider is the image separation of halfway left and halfway right instruments. If standard deviation is 0°, instruments halfway left in the orchestra are reproduced halfway left between the loudspeakers (that is, at 15° off center for speakers separated ±30°). If standard deviation is large, this is the exaggerated separation effect mentioned earlier.

Here are his findings for various fixed mic arrays.

Coincident cardioids at 90°:

The recording angle is ±90° (180° in all). In other words, the orchestra must form a semicircle (180°) around the microphone pair to be reproduced from speaker-to-speaker.

The standard deviation is about 6°. In other words, an instrument that is half-right in the orchestra would be reproduced 6° beyond half-right.

Coincident figure eights at 90° (Blumlein):

Recording angle is ±45° (90° in all).

Standard deviation is about 5°.

Cardioids angled 110° and spaced 17 cm (ORTF):

Recording angle is ±50° (100° in all).

Standard deviation is about 5°.

Cardioids angled 90° and spaced 30 cm (NOS):

Recording angle is ±40° (80° in all).

Standard deviation is about 4°.

Omnis spaced 50 cm (20 in.):

Recording angle is ±50°(100° in all).

Standard deviation is about 8°.

Williams's paper has graphs showing the calculated recording angle and standard deviation for a wide range of polar patterns, anglings, and spacings, as well as other useful information.

Carl Ceoen, "Comparative Stereophonic Listening Tests" (1972)

Carl Ceoen used listening tests to compare several typical stereo techniques. He reported the following average resolution distortion (image focus or sharpness) for these methods:

XY (coincident cardioids angled 135°): 3°

MS (equivalent to coincident hypercardioids angled apart): 5.5°

Blumlein (coincident bidirectionals angled 90°): 4°

ORTF (coincident cardioids at 110°, 17 cm): 3°

NOS (cardioids angled 90° and spaced 30 cm): 4°

Pan-pot: 3°

According to Ceoen, the listening audience agreed that the ORTF system was the best overall compromise and that the MS system lacked intimacy.

Benjamin Bernfeld and Bennett Smith, "Computer-Aided Model of Stereophonic Systems" (1978)

Bernfeld and Smith computed the image location versus frequency for various stereo-miking techniques. The better the coincidence of image locations at various frequencies, the sharper is the imaging. Here are the condensed results:

Blumlein (coincident bidirectionals at 90°): Image focus is good except near the speakers; there, high frequencies are reproduced with a wider stereo spread than low frequencies.

Coincident cardioids angled 90° apart: Image focus is very good, but the stereo spread is very narrow.

Coincident cardioids angled 120° apart: Image focus is fairly good, but the stereo spread is narrow.

Coincident hypercardioids angled 120° apart: Image focus is good but not excellent because high frequencies around 3 kHz are reproduced with a wider spread than low frequencies.

Coincident hypercardioids angled 120° apart, compensated with Vanderlyn's shuffler circuit (Vanderlyn, 1954): Excellent image focus and stereo spread.

Blumlein (coincident bidirectionals at 90°), compensated with shuffler circuit: Very good image focus and stereo spread.

ORTF (cardioids angled 110° and spaced 17 cm): Good image focus; low frequencies have narrow spread and high frequencies have wide spread.

ORTF with hypercardioids: Similar to the above, with wider stereo separation.

Two omnis spaced 9.5 feet: Poor image focus; high frequencies have much wider spread than low frequencies; exaggerated separation effect.

Three cardioids spaced 5 feet: Poor image focus as above, with exaggerated separation at high frequencies.

C. Huggonet and J. Jouhaneau, "Comparative Spatial Transfer Function of Six Different Stereophonic Systems" (1987)

Huggonet and Jouhaneau used a modulated tone burst at various frequencies, plus a violin, with listening tests to compare the spatial transfer function of six different stereophonic systems. Each system has an angular dispersion (image spread) that depends on frequency. In general, the angular dispersion of coincident systems was least. The Blumlein array gave the sharpest imaging, the dummy head and the NOS system the worst. The dummy head gave the best depth perception, followed by ORTF. MS gave the worst depth perception.

M. Hibbing, "XY and MS Microphone Techniques in Comparison" (1989)

In comparing XY and MS coincident methods, Hibbing concluded that MS has several advantages over XY:

1. The MS system can use an omnidirectional mid element, but the XY system cannot use omnidirectional capsules. Since an omni capsule generally has better low-frequency response than a uni, the MS system can have better low-frequency response than the XY system.
2. With MS, any stereo spread can be had with any polar pattern. XY is more limited.
3. With MS, a wider source angle is usable than with XY if polar patterns with a low bidirectional component are used.
4. With MS, the mid element aims at the center of the sound source, so most of the sound arrives close to on axis. With XY, most of the sound arrives off axis and is subject to off-axis coloration.
5. With MS, both the mid and side polar patterns are more uniform with frequency than the patterns in the XY configuration. Consequently, the left/right polar patterns generated by MS are more uniform with frequency than those of XY.

6. With MS, the stereo spread is easy to control by a fader. With XY, the stereo spread must be adjusted mechanically. MS allows stereo-spread adjustment after the session; XY does not.
7. With MS, the mid (mono sum) signal is independent of the stereo spread, so it stays consistent and predictable. With XY, the mono sum varies with the angle between the microphones.

Wieslaw Woszcyk, "A New Method for Spatial Enhancement in Stereo and Surround Recording" (1990)

Using female speech and a soprano recorder as a sound source, Woszcyk recorded the source with several stereo arrays in the diffuse field (29–80 feet from the source). Blind listening tests were done using a stereo pair of speakers and a Dolby surround system. The latter system used front left and right speakers, a center front speaker, a center rear speaker, and left/right surround speakers.

The results are briefly summarized. In general, listening in Dolby surround reduces the stereo separation (stage width) because of the center speaker. Mic techniques for Dolby surround should be optimized to counteract this effect.

In these descriptions, *stage width* means the perceived width of the stage (about ±65° in front of the mic pair). *Spatial effect* means the perceived spaciousness of the concert hall.

XY at 90°: Very narrow stage width, narrow spatial effect.

XY at 180°: Extremely wide stage width and weak center image in stereo, but fairly accurate in surround. This method gave the best spatial effect of the listening test: wide, intense, and natural.

ORTF with cardioids: Fairly accurate stage width in stereo but much narrower in surround. Narrow spatial effect.

ORTF with hypercardioids: Wide stage width, "split" spatial effect.

Blumlein: Accurate stage width up to ±45° in stereo, slightly narrower in Dolby surround. Wide and smooth spatial effect.

14" spaced omni pair: Somewhat narrow stage width in stereo, even less in surround. Smooth and natural spatial effect.

Dummy head: Wide stage width in stereo, narrower in surround. Superior spatial effect: wide and smooth.

PZM wedge (two 18" x 29" hard baffles angled 45°): Overly wide stage width in stereo but accurate in surround. Superior, natural spatial effect. Somewhat honky coloration. Spherical microphones should produce equally good imaging but without coloration.

Summary

Although these experimenters disagree in certain areas, they all agree that widely spaced microphones give poorly focused imaging and that the Blumlein technique gives sharp imaging. Blumlein and Bernfeld say that the imaging of the Blumlein array can be further sharpened with a shuffler or spatial equalizer. Ceoen's results indicate that ORTF is best, but others report less-than-optimal image focus with ORTF.

The most accurate systems for frontal stereo appear to be coincident or near-coincident arrays with spatial equalization, or dual MS arrays. The near-coincident/spaced-pair hybrid method used by Delos also works quite well.

It helps to know about all the stereo techniques to conquer the acoustic problems of various halls or to create specific effects. No particular technique is magic; you often can improve the results by changing the microphone angling or spacing.

References

B. Bartlett. "Stereo Microphone Technique." *db.* 13, no. 12 (December 1979), pp. 310–346.

B. Bernfeld and B. Smith. "Computer-Aided Model of Stereophonic Systems." Audio Engineering Society Preprint 1321, 59th Convention, 1978-02, p. 14.

A. Blumlein. British patent specification. *Journal of the Audio Engineering Society* 6, no. 2 (April 1958), p. 91. Also in *Stereophonic Techniques Anthology.* New York: Audio Engineering Society, 1986.

C. Ceoen. "Comparative Stereophonic Listening Tests." *Journal of the Audio Engineering Society* 20, no. 1 (January–February 1972), pp. 19–27. Also in the *Stereophonic Techniques Anthology.* New York: Audio Engineering Society, 1986.

R. Condamines. "La Prise De Son." In *Stereophonic.* Paris and New York: Masson Publishers, 1978.

K. Farrar. "Sound Field Microphone.," *Wireless World* (October 1979).

T. Faulkner. "Phased Array Recording." *The Audio Amateur* (January 1982).

M. Gayford. *Microphone Engineering Handbook.* Trowbridge, Wiltshire: Focal Press, 1994.

H. Gerlach. "Stereo Sound Recording with Shotgun Microphones." *Journal of the Audio Engineering Society* 37, no. 10 (October 1989), pp. 832–838.

M. Gerzon. "Blumlein Stereo Microphone Technique." *Journal of the Audio Engineering Society* 24, no. 11 (January–February 1976), p. 36.

D. Griesinger. "Spaciousness and Localization in Listening Rooms and Their Effects on Recording Technique." *Journal of the Audio Engineering Society* 34, no. 4 (April 1986), pp. 255–268.

D. Griesinger. "New Perspectives on Coincident and Semi-Coincident Microphone Arrays." Preprint No. 2464 (H4), paper presented at the Audio Engineering Society 82nd convention, March 10–13, 1987, London.

M. Hibbing. "XY and MS Microphone Techniques in Comparison." *Journal of the Audio Engineering Society* 37, no. 10 (October 1989), pp. 823–831.

C. Huggonet and J. Jouhaneau. "Comparative Spatial Transfer Function of Six Different Stereophonic Systems." Preprint 2465 (H5), paper presented at the Audio Engineering Society 82nd convention, March 10–13, 1987, London.

J. Jecklin. "A Different Way to Record Classical Music." *Journal of the Audio Engineering Society* 29, no. 5 (May 1981), pp. 329–332. Also in *Stereophonic Techniques Anthology.* New York: Audio Engineering Society, 1986.

J. Lemon. "Spacing for Fidelity," letter to the editor. *Recording Engineer/Producer* (September 1989), p. 76.

S. Lipshitz. "Stereo Microphone Techniques: Are the Purists Wrong?" *Journal of the Audio Engineering Society* 34, no. 9 (September 1986), pp. 716–744.

L. Olson. "The Stereo-180 Microphone System." *Journal of the Audio Engineering Society* 27, no. 3 (March 1979), pp. 158–163. Also in *Stereophonic Techniques Anthology*. New York: Audio Engineering Society, 1986.

S. Pizzi. "Stereo Microphone Techniques for Broadcast." Preprint No. 2146 (D-3), paper presented at the Audio Engineering Society 76th convention, October 8–11, 1984, New York.

R. Streicher and W. Dooley. "Basic Stereo Microphone Perspectives—A Review." *Journal of the Audio Engineering Society* 33, nos. 7–8 (July–August 1985), pp. 548–556. Also in *Stereophonic Techniques Anthology.* New York: Audio Engineering Society, 1986.

The Symphonic Sound Stage, vol. 2. Delos compact disc D/CD 3504.

P. Vanderlyn, British patent specification 23989 (1954).

M. Williams. "Unified Theory of Microphone Systems or Stereophonic Sound Recording." Preprint No. 2466 (H-6), paper presented at the Audio Engineering Society 82nd convention, March 10–13, 1987, London.

W. Woszcyk. "A New Method for Spatial Enhancement in Stereo and Surround Recording." Preprint 2946, paper presented at the Audio Engineering Society 89th convention, September 21–25, 1990, Los Angeles.

In 1991, a mid-side computer program by John Woram became available from Gotham Audio. You enter the mid pattern and mid/side ratio, then the program displays the resulting left-right polar patterns, pickup angle, and other useful information. (Gotham Audio Corp., 1790 Broadway, New York, NY 10019-1412. Tel. (212) 765-3410.)

11

STEREO BOUNDARY-MICROPHONE ARRAYS

Boundary microphones (discussed in Chapter 7) can make excellent stereo recordings. This chapter explains the characteristics of several boundary-mic arrays.

First, we look at ways to create basic stereo arrays using boundary microphones.

- To make a spaced-pair boundary array, space two boundary microphones a few feet apart. Place them on the floor, a wall, or stand-mounted panels.
- To make a coincident array, mount two boundary mics back to back on a large panel, with the edge of the panel aiming at the sound source.
- To make a near-coincident array, mount each boundary microphone on a separate panel, and angle the panels apart. Or use two directional boundary mics on the floor, angled and spaced.

Boundary microphones can be placed directly on the floor or can be raised above it. We explain several stereo techniques using both methods.

Techniques Using Floor-Mounted Mics

You can place two boundary microphones on the floor to record in stereo. Floor mounting provides several advantages:

- Phase cancellations due to sound reflections off the floor are eliminated.
- Floor mounting provides the best low-frequency response for boundary microphones.
- The mics are very easy to place.
- The mics are nearly invisible. At live concerts, hiding the microphones is often the main consideration.

When a floor-placed boundary array is used to record an orchestra, the front-row musicians usually are reproduced too loudly, due to their relative proximity to the microphones. Musical groups with little front-to-back depth—such as small chamber groups, jazz groups, or soloists—may be the best application for this system.

Let's consider specific techniques for floor-mounted microphones.

Floor-Mounted Boundary Microphones Spaced 4 Feet Apart

Listening tests showed that a spacing of 3–4 feet between microphones is sufficient for a full stereo spread, when the sides of the musical ensemble are 45° off center, from the viewpoint of the center of the microphone array (see Figure 11–1).

With a floor-placed array, the stereo spread decreases as the sound-source height increases. For example, if you record a group of people standing, the spread will be narrower than if the people were sitting. That is because the higher the source is, the less is the time difference between microphones.

Spaced boundary mics have the same drawbacks as spaced conventional mics: poorly focused images, potential lack of mono compatibility, and large phase differences between channels.

Two advantages, however, are a warm sense of ambience and a good stereo effect even for off-center listeners. And, with a spaced pair, you can use omnidirectional boundary microphones without plexiglass boundaries. Therefore, the low-frequency response is excellent and the mics are inconspicuous.

Floor-Mounted Directional Boundary Microphones

Two of these mics can be set up as a near-coincident pair or a spaced pair. For near-coincident use, place the mics on the floor side by side and angle them apart (Figure 11–2). Adjust their angling and spacing for the desired

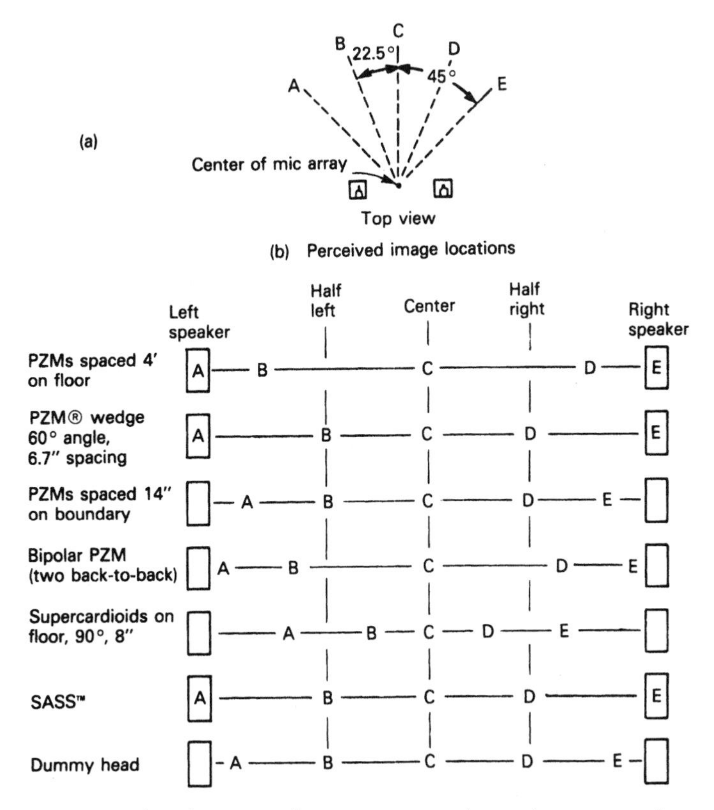

Figure 11–1 Stereo localization of various stereo boundary-microphone arrays: (a) the letters A through E are live speech-source positions relative to mic array, (b) stereo image localization of various stereo mic arrays (listener's perception). Images A through E correspond to live speech sources A through E in (a).

stereo spread. This is an effective arrangement for recording stage plays or musicals. Other mics will be needed for the pit orchestra.

As shown in Figure 11–1, a floor-mounted array of supercardioid boundary mics, angled 90° and spaced 8 inches apart, provides a narrow stage width.

More spacing or angling is needed for accurate localization. The image focus is sharper with this arrangement than with spaced boundary microphones.

L^2 (Lamm-Lehmann) Floor Array

This is a stereo PZM array designed by Mike Lamm and John Lehmann of Dove & Note Recording, Houston, Texas (Bartlett, 1999). It is not

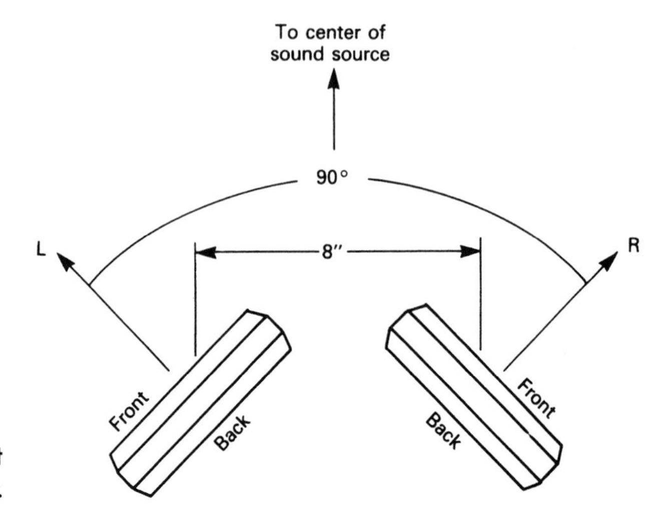

Figure 11–2
Floor-mounted directional boundary microphones set in a near-coincident array.

commercially available, but you can build one as shown in Figure 11–3. The boundaries create directional polar patterns that are angled apart and ear spaced.

According to one user, "You can take this array, set it down, and just roll. You get a very close approximation of the real event."

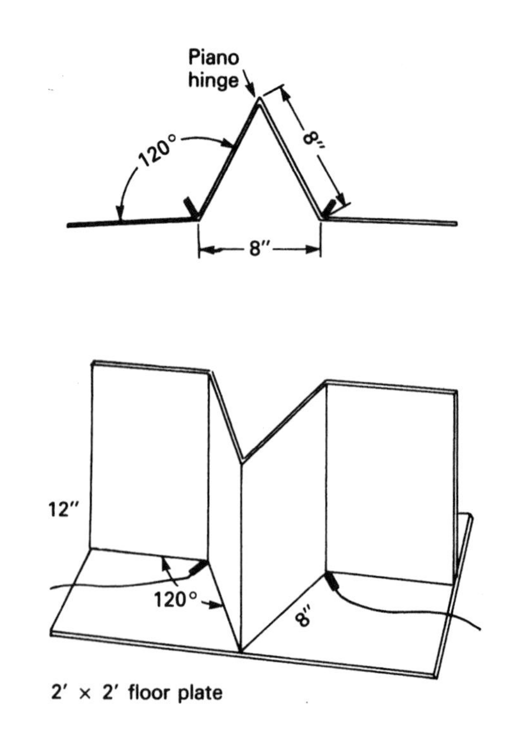

Figure 11–3
L^2 floor array.

Suspending the inverted array from cables results in less bass and more highs, while placing it on the floor reverses this tonal balance.

OSS Boundary-Microphone Floor Array

In this configuration by Josephson Engineering (1988), two boundary microphones are on opposite sides of a hard, absorbent baffle or Jecklin disk cut in half. This array has the characteristics of the OSS system described in the previous chapter plus the advantages of boundary miking.

Floor-Mounted Boundary Microphones Configured for MS

The MS technique can be applied to boundary microphones. The following method was invented by Jerry Bruck (1985) of Posthorn Recordings. The mid unit is an omnidirectional boundary microphone; the side unit is a small-diameter bidirectional condenser microphone mounted a few millimeters above the omni unit.

The bidirectional microphone is close enough to the boundary to prevent phase cancellations between direct and reflected sounds over most of the audible spectrum. Bruck proposed other systems using three transducers.

Since the mid microphone is a boundary unit, it has the same high-frequency response anywhere around it (no off-axis coloration). This contributes to very sharp stereo imaging. And, since the mid capsule is an omni condenser unit, it has excellent low-frequency response. The system is low in profile and unobtrusive.

Like other floor-mounted arrays, this system is limited to recording small ensembles or soloists. It also could be used on a piano lid. No microphones are made this way; you must set a bidirectional microphone over a boundary microphone to form the array.

Techniques Using Raised-Boundary Mics

Some stereo microphone arrays use directional microphones. How do we make an omnidirectional boundary microphone directional? Mount it on a panel (boundary). Then raise the mic or panel several feet off the floor to record large ensembles. The panel makes the microphone reject sounds coming from behind the boundary. For sounds approaching the rear of the panel, low frequencies are rejected least and high frequencies are rejected most.

A small boundary makes the microphone directional only at high frequencies. Low frequencies diffract or bend around a small boundary as if it weren't there. The bigger you make the boundary assembly, the more directional the microphone will be across the audible band.

The bigger the boundary, the lower is the frequency at which the microphone becomes directional. A microphone on a square panel is omnidirectional at very low frequencies and starts to become directional above the frequency *F*, where

$$F = 188/D$$

D = the boundary dimension in feet

Boundaries create different polar patterns at different frequencies. For example, a 2-foot-square panel is omnidirectional at and below 94 Hz. At mid frequencies, the polar pattern becomes supercardioid. At high frequencies, the polar pattern approaches a hemisphere (as in Figure 11–4) (Bartlett, 1999). (Polar patterns are covered in Chapter 7.)

PZM Wedge

A popular boundary configuration for stereo is the PZM wedge (Bartlett, 1999). You start with two clear plexiglass panels about 2 feet square. Join their edges with a hinge or tape to form a V. Tape a PZM to each panel, as shown in Figure 11–5, and aim the point of the V at the sound source. This forms a near-coincident array that has very sharp imaging and accurate localization (Defossez, 1986). It also is mono compatible to a large degree. The angle between boundaries can be varied to change the direct/reverb ratio. The wider the angle, the more forward is the directionality and the closer the source sounds.

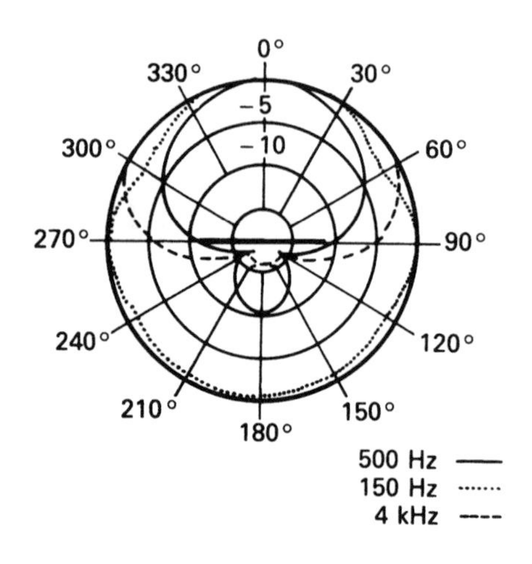

Figure 11–4
Polar patterns at various frequencies on a 2-foot-square panel. (Courtesy of Crown International.)

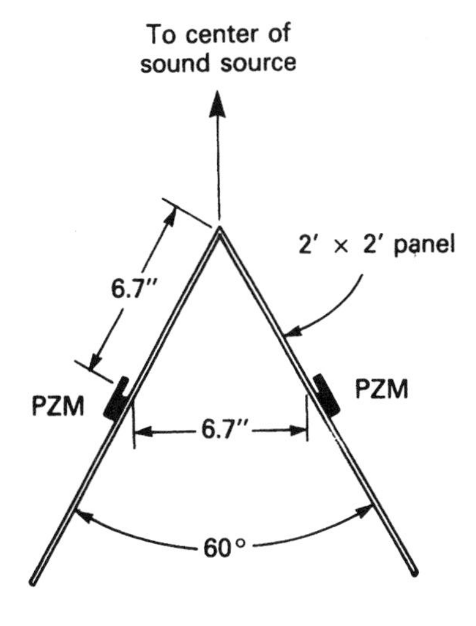

Figure 11–5
PZM wedge.

L^2 Array

Shown in Figure 11–6, this multipurpose array was designed by Mike Lamm and John Lehmann of Dove and Note Recording in Houston, Texas.

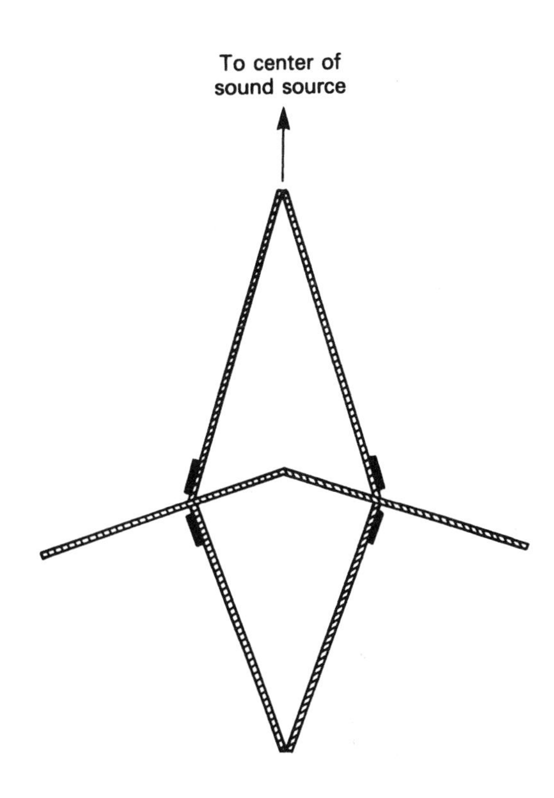

Figure 11–6
L^2 array.

Although it is not commercially available, you can build one as suggested in Lamm and Lehmann (1983). Mike used this array extensively for overall stereo or quad pickup of large musical ensembles. The hinged, sliding panels can be adjusted to obtain almost any stereo pickup pattern.

Pillon PZM Stereo Shotgun Array

This stereo PZM array was devised by Gary Pillon, a sound mixer at General Television Network of Oak Park, Michigan (Bartlett, 1999). Each PZM capsule is in the apex of a pyramid-shaped boundary structure; this produces a highly directional polar pattern. The device is not commercially available, but you can build one as shown in Figure 11–7. The assembly can be stand-mounted from the backside or handheld if necessary. The stereo imaging, which is partly a result of the 8 inch capsule spacing, is designed to be like that produced by a binaural recording but with more realistic playback over loudspeakers. Ideally, this device would mount on a Steadicam platform and give an excellent match between audio and video perspectives.

The Stereo Ambient Sampling System

A stereo microphone has been developed that is purported to solve many problems of other stereo-miking methods. Called the *stereo ambient sampling system* (SASS), it is a stereo condenser microphone using boundary-microphone technology (Bartlett, 1989; Billingsley, 1987, 1989a, 1989b, 1989c). SASS is designed to give highly localized stereo imaging for loud-

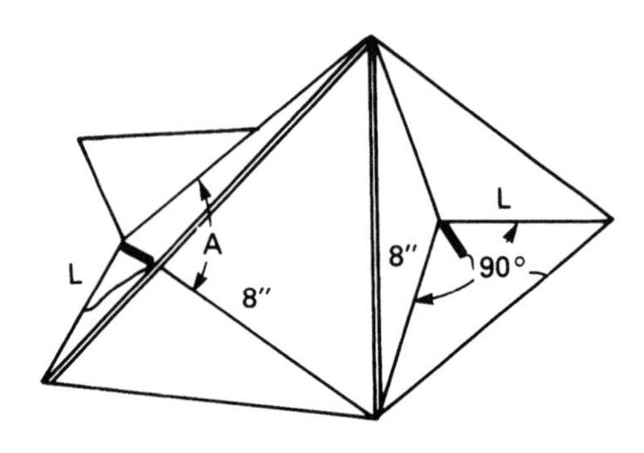

Figure 11–7
Pillon PZM stereo shotgun array.

speaker or headphone reproduction. The device is a mono-compatible, near-coincident array.

Since the microphone has an unusual design and is an example of recent technology, it requires some discussion.

SASS Construction

One model uses two high-quality Pressure Zone Microphones mounted on boundaries to make each microphone directional (as shown in Figure 11–8) Another model (now discontinued) is similar but uses two flush-mounted Bruel & Kjaer 4006 microphones for 10 dB lower noise.

For each channel, an omnidirectional microphone capsule is mounted very near or flush with a boundary approximately 12.7 cm (5 inches) square. The two boundaries are angled left and right of center. The sound diffraction of each boundary, in conjunction with a foam barrier between the capsules, creates a directional polar pattern at high frequencies. The patterns aim left and right of center, much like a near-coincident array. The capsules are "ear spaced" 17 cm (6.7 inches) apart.

The polar patterns of the boundaries and the spacing between capsules have been chosen to provide natural perceived stereo imaging.

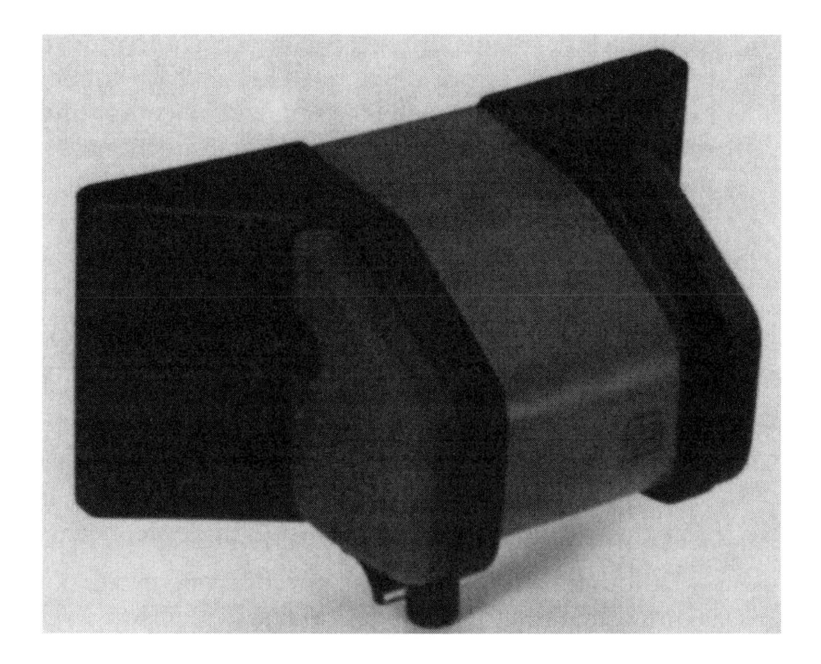

Figure 11–8 Crown SASS-P MKII PZM stereo microphone. (Courtesy of Crown International, Inc.)

The foam barrier or baffle between the capsules limits acoustic crosstalk between the two sides at higher frequencies. Although the microphone capsules are spaced apart, little phase cancellation occurs when both channels are combined to mono because of the shadowing effect of the baffle. That is, despite phase differences between channels, the extreme amplitude differences (caused by the baffle) reduce phase cancellations in mono.

SASS Frequency Response

You might expect the SASS to have poor low-frequency response because it has small boundaries. However, it still has a flat response down to low frequencies. There is no 6 dB shelf because, since the capsules are omnidirectional below 500 Hz, their output at low frequencies is equal in level. The equal-level outputs from the capsules are summed in stereo listening, which causes a 3 dB rise in perceived level at low frequencies. This effectively counteracts 3 dB of the 6 dB low-frequency shelf normally experienced with small boundaries.

In addition, when the microphone is used in a reverberant sound field, the effective low-frequency level is boosted another 3 dB because the pattern is omnidirectional at low frequencies and unidirectional at high frequencies.

All the low-frequency shelf is compensated, so the effective frequency response is uniform from 20 Hz to 20 kHz. According to the manufacturer, this can be proven in an A-B listening test by comparing the tonal balance of the SASS to that of flat-response omnidirectional microphones. They sound tonally the same at low frequencies.

SASS Localization Mechanisms

Like an artificial head (described in the next chapter), the SASS localizes images by time and spectral differences between channels. The localization mechanism varies with frequency:

- Below 500 Hz, the SASS picks up sounds equally in both channels but with a direction-dependent delay between channels.
- Above 500 Hz, SASS localization is due to a combination of time and intensity differences. The intensity difference increases with frequency.

Although this is the opposite of the Cooper-Bauck theoretical criteria for natural imaging over loudspeakers, it seems to work well in practice. Also, it is very close to the mechanism used for binaural recording and by the human hearing system. As stated in Chapter 9, Theile suggests that head-related signals are the best for stereo reproduction.

Listening tests have shown that the SASS does not localize fundamentals with a smoothly rising and falling envelope below 261 Hz. However, the SASS uses the ears' natural ability to localize to the harmonics of a source and ignore the fundamentals. Since the SASS uses delay panning only at low frequencies, its low-frequency localization over loudspeakers might be improved by a Blumlein-type shuffler that Blumlein suggested for closely spaced omnis.

According to David Griesinger (1987), delay panning does not work on lowpass-filtered male speech below 500 Hz. However, in *Spatial Hearing* (1983, pp. 206–207), Jens Blauert shows evidence by Wendt that delay panning does work at 327 Hz. The effect may depend on the shape of the signal envelope, individual listening ability, and training.

Extensive application notes for the SASS are given in Bartlett (1989) and Billingsley, 1987, 1989a, 1989b, 1989c, 1990).

SASS Advantages and Disadvantages

The SASS is claimed to have several characteristics that make it superior to conventional stereo microphone arrays:

- Compared to a coincident pair, the SASS has better low-frequency response and more "air," or spaciousness.
- Compared to a stereo microphone, the SASS costs much less but is relatively large.
- Compared to a near-coincident pair, the SASS has better low-frequency response. Also, it is equally mono-compatible. When you talk into the SASS from all directions in a reverberant room, it sounds tonally the same whether you listen in stereo or mono. That is, it has approximately the same frequency response in mono and stereo.
- Compared to a spaced pair, the SASS has much sharper imaging and less phase difference between channels, making it easier to cut records from SASS recordings. Also, the SASS on a single mic stand is smaller and easier to position than conventional mics on two or three stands.
- Compared to an artificial head used for binaural recording, the SASS is less conspicuous, provides a much flatter response without equalization, provides some forward directionality, and is more mono-compatible. Its binaural localization is not quite as good as that of the artificial head.

The SASS could be called a *quasi-binaural system,* in that it uses many of the same localization mechanisms as an artificial head. A full explanation of binaural recording is provided in the next chapter.

Sphere Microphones

A sphere microphone uses a hard globe 8 inches in diameter, with a pair of pressure-response omni mic capsules flush mounted in either side, 180° apart. Two examples are the Neumann KFM 100 and the Schoeps KFM 6.

Like the SASS, a spherical stereo mic uses time and spectral differences between channels to create stereo images. A circuit corrects the frequency response and phase response of the capsules in the sphere.

Claimed benefits are accurate and sharp imaging, excellent reproduction of depth, extended low-frequency response, and low pickup of wind and vibration. The sphere shape is used because it provides the least diffraction (disturbance of the sound field). As a result, the frequency response is flat not only for sounds in front of the sphere but also for reverberant, diffuse sound.

The mic is largely mono-compatible because, in the bass frequencies, phase shift is small and, in the high frequencies, the acoustic shadow of the sphere produces strong interchannel differences making phase cancellations less probable.

A sphere mic is not the same as a dummy head. Sphere-mic recordings are for speaker listening; dummy-head recordings are for headphone listening.

References

B. Bartlett. *Crown Boundary Microphone Application Guide.* Elkhart, IN: Crown International, 1999.

B. Bartlett. "An Improved Stereo Microphone Array Using Boundary Technology: Theoretical Aspects." Preprint No. 2788 (A-1), paper presented at the Audio Engineering Society 86th convention, March 7–10, 1989, Hamburg.

M. Billingsley. U.S. Patent 4,658,931 (April 21, 1987).

M. Billingsley. "Practical Field Recording Applications for An Improved Stereo Microphone Array Using Boundary Technology." Preprint No. 2788 (A-1), paper presented at the Audio Engineering Society 86th convention, March 7–10, 1989a, Hamburg.

M. Billingsley. "An Improved Stereo Microphone Array for Pop Music Recording." Preprint No. 2791 (A-2), paper presented at the Audio Engineering Society 86th convention, March 7–10, 1989b, Hamburg.

M. Billingsley. "A Stereo Microphone for Contemporary Recording." *Recording Engineer/Producer* (November 1989c).

M. Billingsley. "Theory and Application of a New Near-Coincident Stereo Microphone Array for Soundtrack, Special Effects and Ambience." Paper presented at the 89th Audio Engineering Convention, September 21–25, 1990, Los Angeles.

J. Bruck. "The Boundary Layer Mid/Side (M/S) Microphone: A New Tool." Preprint No. 2313 (C-11), paper presented at the Audio Engineering Society 79th convention, October 12–16, 1985, New York.

A. Defossez. "Stereophonic Pickup System Using Baffled Pressure Microphones." Preprint No. 2352 (D4), paper presented at the Audio Engineering Society 80th convention, March 4–7, 1986, Montreux, Switzerland.

D. Griesinger. "New Perspectives on Coincident and Semi-Coincident Microphone Arrays." Preprint No. 2464 (H4), paper presented at the Audio Engineering Society 82nd convention, March 10–13, 1987, London.

Josephson Engineering. Catalog. San Jose, CA: Josephson Engineering, 1988.

M. Lamm and J. Lehmann. "Realistic Stereo Miking for Classical Recording." *Recording Engineer/Producer* (August 1983), pp. 107–109.

I highly recommend the following recording, which demonstrates the imaging differences among various free-field stereo microphone techniques: *The Performance Recordings Demonstration of Stereo Microphone Technique* (PR-6-CD), recorded by James Boyk, Mark Fischman, Greg Jensen, and Bruce Miller; distributed by Harmonia Mundi USA, 3364 S. Robertson Blvd., Los Angeles, CA 90034, tel. (213) 559-0802. It is available in record stores nationwide.

Surprisingly, you'll hear that different transducer types have different imaging. Why? For sharpest imaging, microphone polar patterns and off-axis phase shift should be uniform with frequency. In a ribbon mic, these needs are met. But a condenser mic tends to be less uniform in frequency, and a dynamic tends to be still less uniform. These characteristics affect the imaging of a stereo pair of microphones.

12

BINAURAL AND TRANSAURAL TECHNIQUES

This chapter covers binaural recording with an artificial (dummy) head. The head contains a microphone flush mounted in each ear. You record with these microphones and play back the recording over headphones. This process can re-create the locations of the original performers and their acoustic environment with exciting realism.

Also covered in this chapter is transaural stereo, which is loudspeaker playback of binaural recordings, specially processed to provide surround sound with only two speakers up front.

Binaural Recording and the Artificial Head

Binaural (two-ear) recording starts with an artificial head or dummy head. This is a model of a human head with a flush-mounted microphone in each ear (Figure 12–1). These microphones capture the sound arriving at each ear. The microphones' signals are recorded. When this recording is played back over headphones, your ears hear the signals that originally appeared at the dummy head's ears (Figure 12-2); that is, the original sound at each ear is reproduced (Geil, 1979; Peus, 1989; Sunier, 1989a, 1989b, 1989c; Genuit and Bray, 1989).

Binaural recording works on the following premise. When we listen to a natural sound source in any direction, the input to our ears is just two

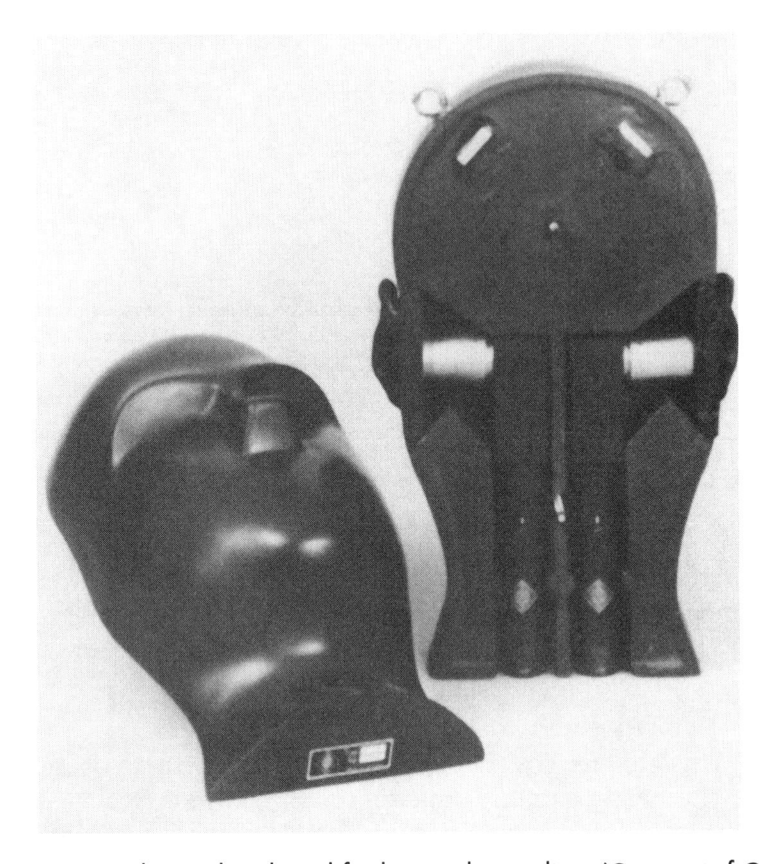

Figure 12–1 A dummy head used for binaural recording. (Courtesy of Gotham Audio Corporation.)

one-dimensional signals: the sound pressures at the ear drums. If we can re-create the same pressures at the listener's eardrums as would have occurred "live," we can reproduce the original listening experience, including directional information and reverberation (Moller, 1989).

Binaural recording with headphone playback is the most spatially accurate method now known. The re-creation of sound-source locations and room ambience is startling. Often, sounds can be reproduced all around your head—in front, behind, above, below, and so on. You may be fooled into thinking that you're hearing a real instrument playing in your listening room.

As for drawbacks, the artificial head is conspicuous, which limits its use for recording live concerts; it is not mono-compatible; and it is relatively expensive.

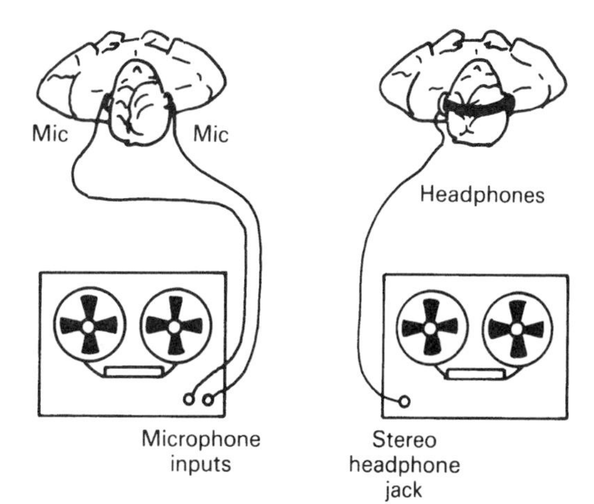

Figure 12–2
Binaural recording and headphone playback.

How It Works

An artificial head picks up sound as a human head does. The head is an obstacle to sound waves at middle to high frequencies. On the side of the head away from the sound source, the ear is in a sonic shadow: The head blocks high frequencies. In contrast, on the side of the head toward the source, there is a pressure buildup (a rise in the frequency response) at middle to high frequencies.

The folds in the pinna (outer ear) also affect the frequency response by reflecting sounds into the ear canal. These reflections combine with the direct sound, causing phase cancellations (dips in the response) at certain frequencies.

The human eardrum is inside the ear canal, which is a resonant tube. The ear canal's resonance does not change with sound-source direction, so the ear canal supplies no localization cues. For this reason, it is omitted in most artificial heads. Typically, the microphone diaphragm is mounted nearly flush with the head, 4 mm inside the ear canal.

To summarize, the head and outer ear cause peaks and dips in the frequency response of the sound received. These peaks and dips vary with the angle of sound incidence; they vary with the sound-source location. The frequency response of an artificial head is different in different directions. In short, the head and outer ear act as a direction-dependent equalizer.

Each ear picks up a different spectrum of amplitude and phase because one ear is shadowed by the head and the ears are spaced apart. These interaural differences vary with the source location around the head.

When the signals from the dummy-head microphones are reproduced over headphones, you hear the same interaural differences that the dummy head picked up. This creates the illusion of images located where the original sources were.

Physically, an artificial head is a near-coincident array using boundary microphones: The head is the boundary, and the microphones are flush mounted in this boundary. The head and outer ears create directional patterns that vary with frequency. The head spaces the microphones about $6\frac{1}{2}$ inches apart. Some dummy heads include shoulders or a torso, which aids front and back localization in human listening but can degrade it in binaural recording and playback (Griesinger 1989).

The microphones in a near-coincident array are directional at all frequencies and use no baffle between them. In contrast, the mics in an artificial head are omni at low frequencies and unidirectional at high frequencies (due to the head baffle effect).

Ideally, the artificial head is as solid as a human head, to attentuate sound passing through it (Sunier, 1989c). For example, the Aachen head is made of molded dense fiberglass (Genuit and Bray, 1989).

You can substitute your own head for the artificial head by placing miniature condenser microphones in your ears and recording with them. The more that a dummy head and ears are shaped like your particular head and ears, the better is the reproduced imaging. Thus, if you record binaurally with your own head, you might experience more precise imaging than if you recorded with a dummy head. This recording will have a nonflat response because of head diffraction (which I explain later).

Another substitute for a dummy head is a head-sized sphere with flush-mounted microphones where the ears would be. This system, called the *Kugelflachenmikrofon,* was developed by Gunther Theile for improved imaging over loudspeakers (Griesinger, 1989).

In-Head Localization

Sometimes the images are heard inside your head rather than outside. One reason has to do with head movement. When you listen to a sound source that is outside your head and move your head slightly, you hear small changes in the arrival-time differences at your ears. This is a cue to the brain that the source is outside your head. Small movements of your head help to externalize sound sources. But the dummy head lacks this cue because it is stationary.

Another reason for in-head localization is that the conch resonance of the pinna is disturbed by most headphones. The conch is the large cavity in the pinna just outside the ear canal. If you equalize the headphone signal to restore the conch resonance, you hear images outside the head (Cooper and Bauck, 1989).

Artificial-Head Equalization

An artificial head (or a human head) has a nonflat frequency response due to the head's diffraction, the disturbance of a sound field by an obstacle. The diffraction of the head and pinnae create a very rough frequency response, generally with a big peak around 3 kHz for frontal sounds. Therefore, binaural recordings sound tonally colored because of this peak unless compensating equalization is used. Some artificial heads have built-in equalization that compensates for the effect of the head.

What is the best equalization for an artificial head to make it sound tonally like a conventional flat-response microphone? Several equalization schemes have been proposed:

- **Diffuse-field equalization:** This compensates for the head's average response to sounds arriving from all directions (such as reverberation in a concert hall).
- **Frontal free-field equalization:** This compensates for the head response to a sound source directly in front, in anechoic conditions.
- **10° averaged, free-field equalization:** This compensates for the head's response to a sound source in anechoic conditions, averaged over ±10° off center.
- **Free-field with source at ±30° equalization:** This compensates for the head's response to a sound source 30° off center, in anechoic conditions. This is a typical stereo loudspeaker location.

The Neumann KU-81i and KEMAR artificial heads use diffuse-field equalization, which Theile also recommends. However, Griesinger (1989) found that the KU-81i needed additional equalization to sound like a Calrec Soundfield microphone: approximately + 7 dB at 3 kHz and + 4 dB at 15 kHz. He prefers either this equalization or a 10° averaged free-field response for artificial heads. The Aachen head, developed by Gierlich and Genuit, is equalized flat for free-field sounds in front (Genuit and Bray, 1989), while Cooper and Bauck (1989) recommend that artificial heads be equalized flat for free-field sounds at ±30°.

To provide a net flat response from microphone to listener, the artificial-head equalization should be the inverse of the headphone frequency response. If the head is equalized with a dip around 3 kHz to yield a net flat response, the headphones should have a mirror-image peak around 3 kHz.

Artificial-Head Imaging with Loudspeakers

How does an artificial-head recording sound when reproduced over loudspeakers? According to Griesinger (1989), it can sound just as good as an ordinary stereo recording, with superior reproduction of location, height, depth, and hall ambience. But it sounds even better over headphones. Images in binaural recordings are mainly up front when you listen with speakers but are all around when you listen with headphones.

Genuit and Bray (1989) report that more reverberation is heard over speakers than over headphones, due to a phenomenon called *binaural reverberance suppression.* For this reason, it is important to monitor artificial-head recordings with headphones and speakers.

Griesinger notes that a dummy head must be placed relatively close to the musical ensemble to yield an adequate ratio of direct-to-reverberant sound over loudspeakers. This placement yields exaggerated stereo separation with a hole in the middle. However, the center image can be made more solid by boosting in the presence range (see Griesinger's, 1989, recommended equalization mentioned previously).

Although a dummy-head binaural recording can provide excellent imaging over headphones, it produces inadequate spaciousness at low frequencies over loudspeakers (Huggonet and Jouhaneau, 1987) unless spatial equalization is used (Griesinger, 1989). (Spatial equalization was discussed in Chapter 3.) A low-frequency boost in the L – R difference signal of about 15 dB at 40 Hz and +1 dB at 400 Hz can improve the low-frequency separation over speakers.

Transaural Stereo

It would be ideal to hear the binaural effect without having to wear headphones. That is, we'd prefer to use loudspeakers to reproduce images all around us. Our ears need only two channels to hear surround sound, so it seems as though we should be able to produce this effect with only two speakers. We can, and this process is called *transaural stereo.*

Transaural stereo converts binaural signals from an artificial head into surround-sound signals played over two loudspeakers. When done correctly, you can hear sounds in any direction around your head, with only two loudspeakers up front (Cooper and Bauck, 1989; Eargle, 1976, Chapters 2 and 3; Bauer, 1961; Schroeder and Atal, 1963; Parsons; Damaske, 1971; Mori et al., 1979; Sakamoto et al., 1978, 1981, 1982; Clegg, 1979; Farrar, 1979; Moller, 1989; Bock and Keele, 1986; Sunier, 1989c). I next explain how it works.

How It Works

When we listen over headphones, the right ear hears only the right signal and the left ear hears only the left signal. But when we listen over loudspeakers, there is acoustic crosstalk around the head (as shown in Figure 12–3). The right ear hears not only the signal from the right speaker but also the signal from the left speaker that travels around the head.

The transaural converter cancels the signal from the left speaker that reaches the right ear and the signal from the right speaker that reaches the left ear. That is, it cancels the acoustic crosstalk from each speaker to the opposite ear, so that the left ear hears only the left speaker and the right ear hears only the right speaker. This is as if you were wearing headphones but without the physical discomfort. The crosstalk-canceling can be applied before or after recording.

Figure 12–4 shows a simplified block diagram of a crosstalk canceler. An equalized, delayed signal is crossfed to the opposite channel and added in opposite polarity. This electronic crosstalk cancels out the acoustic crosstalk that occurs during loudspeaker listening.

Anticrosstalk equalization (EQ) is the difference in frequency response at the two ears due to head diffraction. The delay is the arrival-time difference between ears. The EQ and delay depend on the angle of the speakers off center, which typically is ±30°.

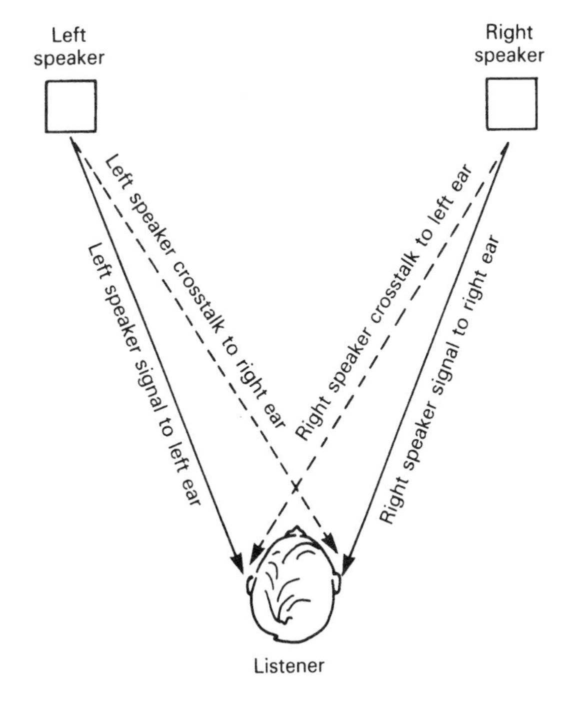

Figure 12–3
Interaural acoustic crosstalk in stereo listening.

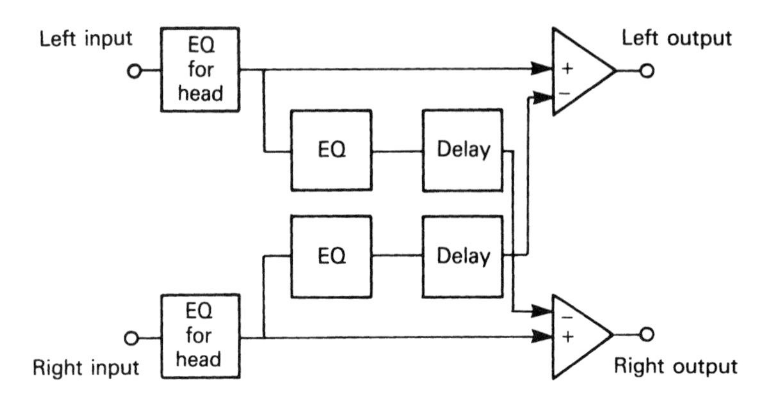

Figure 12–4 A crosstalk canceler.

Figure 12–5 shows how crosstalk cancellation works. Suppose we want to make a left-channel signal appear only at the left ear. That is, we want to cancel the sound from the left speaker that reaches the right ear.

In Figure 12–5,

- L is the direct signal from the left speaker to the left ear.
- R is the head-diffracted (equalized, delayed) signal from the left speaker to the right ear.
- –R is the crosstalk-canceling signal. It is an equalized, delayed, inverted version of signal L.

Signals R and –R add out of phase (cancel) at the right ear, so signal L is heard only at the left ear. Note that the cancellation signal itself is head-diffracted to the opposite ear and also needs to be canceled.

History of Transaural Stereo

Bauer (1961) was the first to suggest such a system. Schroeder and Atal first experimented with transaural stereo in 1962. They used a computer to equalize and crossfeed the two binaural channels and played the result over two speakers in an anechoic chamber. The computer program simulated complicated finite impulse response filters. If the listener moved more than a few inches from the sweet spot for best imaging, the effect was lost. However, Schroeder and Atal reported that the surround effect was "nothing short of amazing" (Schroeder and Atal, 1963; Parsons).

In the late 1970s, JVC demonstrated prototypes of a biphonic processor that produced similar effects. JVC also researched a four-channel system,

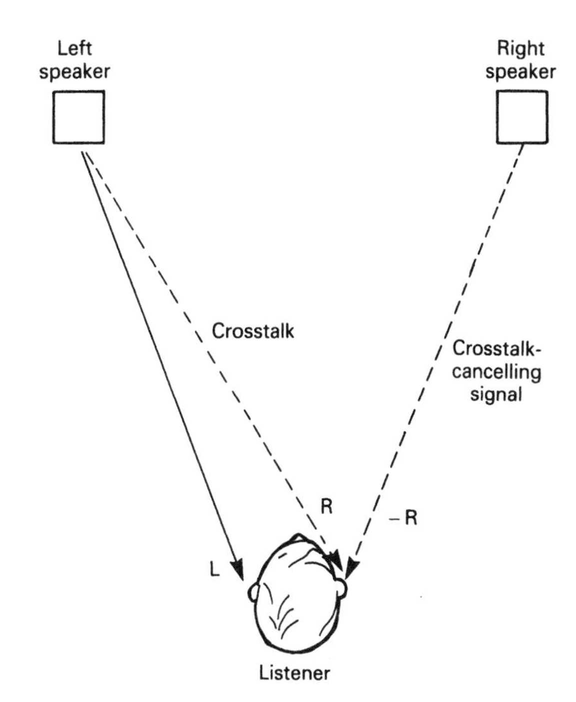

Figure 12–5
Over-the-listener view of crosstalk cancellation.

called *Q-biphonic,* using two dummy heads, one directly in front of the other with a baffle between them (Mori et al., 1979). Research by Matsushita led to a Ramsa sound localization control system with joysticks for surround-sound panning (Sakamoto et al., 1978, 1979, 1980; Clegg, 1979).

Cooper and Bauck's Crosstalk Canceler

Cooper and Bauck (1989) simplified the crosstalk-canceling filters to minimum-phase filters in a shuffler arrangement. This allowed the circuit to be realized in a few op-amps or a digital signal processing chip. They further simplified the filters at high frequencies, which enlarged the sweet spot for good imaging. In addition, they found that anechoic listening is unnecessary: All you need to do is place your speakers 1 or 2 feet from nearby walls to delay early reflections at least 2 milliseconds.

Salava (1990) recommends a close-field speaker setup for transaural imaging. This monitoring arrangement provides the most accurate imaging, nearly independent of the listening room used.

The Cooper and Bauck circuit generates interchannel amplitude differences and phase differences that vary with frequency. These interchannel differences are said to create interaural differences like those produced by real sound sources in various directions.

Basically, their circuit is made of filters and sum-and-difference sections. Filters are necessary to simulate head diffraction, because the spectrum of acoustic crosstalk around the head depends on head diffraction. Two types of filters are used independently: head-diffraction filters and anticrosstalk filters.

The current circuit compensates for crosstalk around a sphere up to 6 kHz. Since the diffraction for a head is different than that for a sphere, you might expect different results depending on the diffraction model used. Usually the crosstalk equalization is based on the head's frequency response to a source 30° off center, where a stereo loudspeaker usually is placed.

Their transaural circuits can be used in many ways:

- As part of a recording system to provide three-dimensional reproduction of music and concert-hall ambience. For recording, the system could use either a dummy head, a spherical stereo mic, or some other system.
- For binaural synthesis, as an inverse shuffler that synthesizes binaural signals from a mono signal. Binaural synthesis is the inverse of crosstalk canceling. An inverse shuffler is a crosstalk filter that mimics the crosstalk occurring when a person listens to stereo speakers. One application of binaural synthesis is as a surround-sound pan pot for recording studios. The engineer can pan a mono signal (say, from a tape track) anywhere around the listener. Another application is for processing conventional microphone signals to simulate transmission around a head to the ears. This creates a binaural recording without having to use a dummy head. A binaural synthesizer has been developed by Gierlich and Genuit (Genuit and Bray, 1989).
- As virtual loudspeakers. These transaural images are synthesized to simulate loudspeakers placed at desired locations. One application is as a speaker spreader for stereo TVs and boom boxes, whose narrow-spaced speakers can be made to sound widely spaced. The circuit could be in either each TV, each broadcast station, or preprocessed recordings. Boers (1983) describes a simpler speaker spreader using antiphase crosstalk. Another application suggested by Cooper and Bauck is a virtual center-channel speaker for theaters.

VMAx

Harman International licenses a transaural process called *VMAx,* based on Cooper and Bauck's transfer function. VMAx simulates 5.1 surround sound with only two speakers. There is virtually no timbral difference among the five phantom speakers.

Transaural licensing is available from Eric Lucas, VP of Technology Licensing, Harman International Industries, Inc., 8500 Balboa Blvd., Northridge, CA 91329, 818-893-8411, e-mail elucas@harman.com.

Medianix sells a chip that implements VMAx, the MED250007 VMAx 3D virtual theater digital audio processor. Medianix can be reached at 415-960-7081.

Lexicon's Transaural Processor

A manufactured product employing transaural principles is the Lexicon CP-3 digital audio environment processor. This consumer-type surround-sound unit can provide transaural processing of binaural recordings (among other programs). Based on work by David Griesinger, it simulates side speakers by crosstalk cancellation; and like Atal and Schroeder's method, the CP-3 cancels the signal that travels around the listener's head and also the signal used for the first cancellation (Sunier, 1989c).

Griesinger reports that the sweet spot for best imaging with the CP-3 is only 2 inches wide, but that the binaural mode "provides the most realistic playback of height, depth, and surround I have yet heard through speakers."

Other Two-Speaker Surround-Sound Systems

- RSP Technologies Circle Surround is a system with a controller, a panel with four joystick panpots, an encoder, and a decoder. The encoder accepts six discrete channels of information and encodes them into two channels. The decoder, used for monitoring, re-creates the six channels from the two coded channels fed to it. Circle Surround is compatible with Dolby Pro Logic and every other type of encoding.
- The Desper spatializer lets you pan tracks almost entirely around the listener, and fore and aft, using only two speakers. Spatializer is a multitrack real-time processor and group of joystick panpots. It is mono-compatible, surround-compatible, and requires no decoding. Spatializer processes the difference (L – R) signal through a mid-boost filter and adds short delays. The product includes MIDI control software, which converts the panpot moves into MIDI data for storage on a sequencer.

- PRO spatializer is a plug-in for the Sonic Solutions editing software, and Spatializer PT3D is a plug-in for Digidesign Pro Tools III. The HTMS-2510 is a playback-only unit for home stereos and surround systems.
- SRS, the sound retrieval system by Arnold Klayman, uses head-related equalization to provide 3-D surround sound from two speakers (and also works with surround decoders). The speakers disappear as sources. There is no sweet spot or critical listening area. SRS can be applied during recording or playback and requires no decoder. It extracts ambience (L – R) information from an audio signal, then applies HRTF (Head-Related Transfer Functions) corrections to create the image locations of the original event. NuReality supplies the SRS processor.
- Dolby Surround Multimedia (DSM) provides a surround-sound experience from two closely spaced computer speakers.
- Crystal River Engineering's Proton Audioreality plug-in for Pro Tools lets you set the position of a sound source in three dimensions. Proton takes a mono signal and gives it the psychoacoustic cues to appear where you want it to, when heard over two speakers. The Alphatron card and software provide surround sound for PC multimedia.
- Q-Sound's QSys/TDS plug-in allows you to position a sound source within a 180° soundstage. It depends on the listener being in the sweet spot of two speakers, when the listener is seated at a multimedia computer.
- The Roland Sound Space (RSS) system allows localization in a full sphere around the listener and image movement in real time, using only two speakers. One model is the RSS-10 with computer control software. It accepts left and right signals and sends them through binaural and transaural processors.
- The ITE/PAR (In The Ear/Pinna Acoustic Response) recording system was developed by Don and Carolyn Davis of Synergetic Audio Concepts. Two high-quality probe mics are inserted into the pressure zone next to the ear drum in the ear canals of a human listener. (In the future, the human listener might be replaced by a dummy head.) The microphones' signals are equalized to have a flat diffuse-field response when mounted in the ears. During playback of the recording, you listen to four speakers, two in front (as for normal stereo) and two on either side, aimed at the ears. The speakers should be placed to reduce early reflections. The side speakers are said to mask opposite-ear crosstalk for a headphonelike effect. Although this is not a

two-speaker system, it creates a surround-sound effect from binaural recordings (*Syn Aud Con Newsletter,* vol. 17, no. 1).

- Virtual surround systems simulate 5.1 surround using only two speakers. Some of these systems are SRS TruSurround, Spatializer N-2-2, Q Surround, Aureal A3D, Dolby Virtual Surround, Sony Virtual Enhanced Surround, Panasonic Virtual Sonic, and VLS Cyclone 3D ("Virtual Surround Systems," 1998).

Although stereo reproduction has been improved thanks to digital recording, it still does not put the listener in the concert hall. The reason is that all the hall reverberation is reproduced up front between the listener's pair of speakers—not all around the listener, as it is in a concert hall. While surround sound can simulate reverb around the listener, it requires extra speakers and power amps. Transaural stereo does the same thing with just two speakers, but it has a relatively small sweet spot of accurate imaging.

References

B. Bauer. "Stereophonic Earphones and Binaural Loudspeakers." *Journal of the Audio Engineering Society* 9, no. 2 (April 1961), pp. 148–151.

T. Bock and D. Keele. "The Effects of Interaural Crosstalk on Stereo Reproduction and Minimizing Interaural Crosstalk in Nearfield Monitoring by the Use of a Physical Barrier, Parts 1 and 2." Preprint Nos. 2420-A (B-10) and 2420-B (B-10), papers presented at the Audio Engineering Society 81st convention, November 12–16, 1986, Los Angeles.

P. Boers. "The Influence of Antiphase Crosstalk on the Localization Cue in Stereo Signals." Preprint No. 1967 (A5), paper presented at the Audio Engineering Society 73rd convention, March 15–18, 1983, Eindhoven, the Netherlands.

A. Clegg. "The Shape of Things to Come: Psycho-Acoustic Space Control Technology." *db* (June 1979), pp. 27–29.

D. Cooper and J. Bauck. "Prospects for Transaural Recording." *Journal of the Audio Engineering Society* 37, nos. 1–2 (January–February 1989), pp. 3–19.

P. Damaske. "Head-Related Two-Channel Stereophony with Loudspeaker Reproduction." *Journal of the Acoustical Society of America* 50, no. 4 (1971), pp. 1109–1115.

J. Eargle. *Sound Recording.* New York: Van Nostrand Reinhold Company, 1976.

K. Farrar. "Sound Field Microphone." *Wireless World* (October 1979).

F. Geil. "Experiments with Binaural Recording." *db* (June 1979), pp. 30–35.

K. Genuit and W. Bray. "The Aachen Head System: Binaural Recording for Headphones and Speakers." *Audio* (December 1989), pp. 58–66.

D. Griesinger. "Equalization and Spatial Equalization of Dummy Head Recordings for Loudspeaker Reproduction." *Journal of the Audio Engineering Society* 37, nos. 1–2 (January–February 1989), pp. 20–29.

C. Huggonet and J. Jouhaneau. "Comparative Spatial Transfer Function of Six Different Stereophonic Systems." Preprint No. 2465 (H5), paper presented at the Audio Engineering Society 82nd convention, March 10–13, 1987, London.

H. Moller. "Reproduction of Artificial-Head Recordings Through Loudspeakers." *Journal of the Audio Engineering Society* 37, nos. 1–2 (January–February 1989), pp. 30–33.

T. Mori, G. Fujiki, N. Takahashi, and F. Maruyama. "Precision Sound-Image-Localization Technique Utilizing Multi-Track Tape Masters." *Journal of the Audio Engineering Society* 27, nos. 1–2 (January–February 1979), pp. 32–38.

T. Parsons. "Super Stereo: Wave of the Future?" *The Audio Amateur.*

S. Peus. "Development of a New Studio Artificial Head." *db* (June 1989), pp. 34–36.

N. Sakamoto, T. Gotoh, T. Kogure, and M. Shimbo. "On the Advanced Stereophonic Reproducing System 'Ambience Stereo.'" Preprint No. 1361 (G3), paper presented at the Audio Engineering Society 60th convention, May 2–5, 1978, Los Angeles.

N. Sakamoto, T. Gotoh, T. Kogure, and M. Shimbo. "Controlling Sound-Image Localization in Stereophonic Reproduction, Part I." *Journal of the Audio Engineering Society* 29, no. 11 (November 1981), pp. 794–799.

N. Sakamoto, T. Gotoh, T. Kogure, and M. Shimbo. "Controlling Sound-Image Localization in Stereophonic Reproduction, Part II." *Journal of the Audio Engineering Society* 30, no. 10 (October 1982), pp. 719–722.

T. Salava. "Transaural Stereo and Near-Field Listening." *Journal of the Audio Engineering Society* 38, nos. 1–2 (January–February 1990), pp. 40–41.

M. Schroeder and B. Atal. "Computer Simulation of Sound Transmission in Rooms." *IEEE Convention Record*, Part 7 (1963), pp. 150–155.

J. Sunier. "A History of Binaural Sound." *Audio* (March 1989a), pp. 312–346.

J. Sunier. "Binaural Overview: Ears Where the Mics Are, Part 1." *Audio* (November 1989b), pp. 75–84.

J. Sunier. "Binaural Overview: Ears Where the Mics Are, Part 2," *Audio* (December 1989c), pp. 48–57.

Syn Aud Con Newsletter 17, no. 1, pp. 12–13. Norman, IN: Syn-Aud-Con.

"Virtual Surround Systems." *Stereo Review* (August 1998).

Several papers on transaural stereo and surround sound were presented at the Audio Engineering Society 89th convention, September 21–25, 1990, Los Angeles:

Durand R. Begault (NASA–Ames Research Center, Moffett Field, CA). "Challenges to the Successful Implementation of 3-D Sound."

W. Bray, K. Genuit, and H. W. Gierlich (Jaffe Acoustics, Norwalk, CT). "Development and Use of Binaural Recording Technology."

Elizabeth A. Cohen (Charles M. Salter Associates, Inc., San Francisco). "Subjective Evaluation of Spatial Image Formation Processors."

D. J. Furlong and A. G. Garvey. "Spaciousness Enhancement of Stereo Reproduction Using Spectral Stereo Techniques," Preprint 3007.

Peter S. Henry (Precision Monolithics, Inc., Santa Clara, CA). "An Analog LSI Dolby TM Pro-Logic Decoder I.C."

Tomlinson Holman (University of Southern California, Los Angeles, and Lucasfilm Ltd., San Rafael, CA). "New Factors in Sound for Cinema and Television."

S. Julstrom. "An Intuitive View of Coincident Stereo Microphones." Preprint 2984.

Gary Kendall and Martin Wilde (Auris Corporation, Evanston, IL). "Spatial Sound Processor for Simulating Natural Acoustic Environments."

Kevin Kotorynski (University of Waterloo, Ontario). "Digital Binaural/Stereo Conversion and Crosstalk Cancelling."

William Martens (Auris Corporation, Evanston, IL). "Directional Perception on the Cone of Confusion."

Gunther Theile (Institut fur Rundfunktechnik, GmbH, Munich). "Multi-Channel Sound in the Home: Further Developments of Stereophony."

Wieslaw R. Woszcyk (McGill University, Montreal). "A New Method for Spatial Enhancement in Stereo and Surround Recording."

More recent Audio Engineering Society preprints include these:

Gregory H. Canfield and Sen M. Kuo (Northern Illinois University, DeKalb). "Dual-Channel Audio Equalization and Cross-Talk Cancellation for Correlated Stereo Signals." Preprint 4570 (J-6), September 1997.

Andre L. G. Defossez (Brussels). "A Matrixed Pressure Triplet for Full Surround Stereophonic Sampling." Preprint 3156 (R-5), October 1991.

K. Genuit, H.W. Gierlich, and Wade Bray (HEAD Acoustics, Aachen, Germany, and Norwalk, CT). "Development and Use of Binaural Recording Techniques." Preprint 2950, September 1990.

K. Genuit et al. (HEAD Acoustics, Herzogenrath, Germany). "Improved Possibilities of Binaural Recording and Playback Techniques." Preprint 3332, March 1992.

Michael Gerzon (Oxford, England). "The Design of Distance Panpots." Preprint 3308, March 1992.

Michael Gerzon (Oxford, England). "Microphone Techniques for Three-Channel Stereo." Preprint 3450, October 1992.

Michael Gerzon (Oxford, England). "Signal Processing for Simulating Realistic Stereo Images." Preprint 3423 (O-1), October 1992.

Michael Gerzon (Oxford, England). "Applications of Blumlein Shuffling to Stereo Microphone Techniques." Preprint 3448 (S-1), October 1992.

Dorte Hammershoi et al. (Institute for Electronic Systems, Aalborg University, Denmark). "Head Related Transfer Functions: Measurements of 24 Human Subjects." Preprint 3289, March 1992.

Yuvi Kahana, Philip A. Nelson, and Ole Kirkeby (University of Southampton) and Hareo Hamada (Tokyo Denki University). "Objective and Subjective Assessment of Systems for the Production of Virtual Acoustic Images for Multiple Listeners." Preprint 4573 (J-9), September 1997.

Ole Kirkeby and Philip A. Nelson (University of Southampton) and Hareo Hamada (Tokyo Denki University). "Virtual Source Imaging Using the 'Stereo Dipole.'" Preprint 4574 (J-10), September 1997.

Teruji Kobayashi (Nittobo Acoustic Engineering Co., Ltd.) and Hareo Hamada and Tanetoshi Miura (Tokyo Denki University). "Binaural Based 3-D Audio System: The Orthostereophonic System (OSS) and Applications." Preprint 3729, October 1993.

Robert C. Maher (EuPhonics, Inc., Boulder, CO). "A Low Complexity Spatial Localization System." Preprint 4567 (J-3), September 1997.

Henrik Moller et al. (Institute for Electronic Systems). "Transfer Characteristics of Headphones." Preprint 3290, March 1992.

Akita Morita and Toshiro Haraga (NHK Science and Technical Research Laboratories, Tokyo) and Keishi Imanaga (Sanken Microphone Co. Ltd., Tokyo). "A 3-1 Quadraphonic Microphone for HDTV." Preprint 3451 (S-4), October 1992.

F. Richter (AKG Acoustics). "BAP Binaural Audio Processor." Preprint 3323, March 1992.

Albert G. Swanson (Location Recording, Seattle). "Standard Stereo Recording Techniques in Non-Standard Situations." Preprint 3313, March 1992.

Gunther Theile (Institut fur Rundfunktechnik GmbH, Munich). "Further Developments of Loudspeaker Stereophony." Preprint 2947, September 1990.

Floyd E. Toole (National Research Council Canada, Ottawa, Ontario). "Binaural Record/Reproduction Systems and Their Use in Psychoacoustic Investigations." Preprint 3179 (L6), October 1991.

Soren Gert Weinrich (Oticon A/S Research Unit, Snekkersten, Denmark). "Improved Externalization and Frontal Perception of Headphone Signals." Preprint 3291, March 1992.

Martin D. Wilde (Wilde Acoustics, Chicago). "Temporal Localization Cues and Their Role in Auditory Perception." Preprint 3708, October 1993.

Michael Williams (Paris). "Early Reflections and Reverberant Field Distribution in Dual Microphone Stereophonic Sound Recording Systems." Preprint 3155 (R-4), October 1991.

Michael Williams (Paris). "Frequency Dependent Hybrid Microphone Arrays for Stereophonic Sound Recording." Preprint 3252, March 1992.

W. Woszczyk (McGill University, Montreal). "Microphone Arrays Optimized for Music Recording." Preprint 3255, March 1992.

Ning Xiang, Klaus Genuit, and Hans W. Gierlich (Head Acoustics, Herzogenrath, Germany). "Investigations on a New Reproduction Procedure for Binaural Recordings." Preprint 3732, October 1993.

Preprints can be ordered for $5.00 each from the Audio Engineering Society, 60 East 42nd St., New York, NY 10165.

Here are more articles from the *Journal of the Audio Engineering Society:*

Durand Begault (Aerospace Human Factors Division, NASA–Ames Research Center, Moffett Field, CA). "Challenges to the Successful Implementation of 3-D Sound." November 1991.

H. L. Han."Measuring a Dummy Head in Search of Pinna Cues." January–February 1994.

F. R. Heegaard (Statsradiofonien of Denmark). "The Reproduction of Sound in Auditory Perspective and a Compatible System of Stereophony." October 1992.

Stanley Lipshitz (Audio Research Group, University of Waterloo, Ontario). "Comments on 'Spaciousness and Localization in Listening Rooms and Their Effects on the Recording Technique.'" December 1987.

E. Macpherson (Audio Research Group, University of Waterloo, Ontario). "A Computer Model of Binaural Localization for Stereo Imaging Measurement." September 1991.

Henrik Moller et al. (Acoustics Laboratory, Aalborg University, Denmark). "Binaural Technique: Do We Need Individual Recordings?" June 1996.

M. Poletti (Audio Signal Processing Group, Industrial Research Ltd., Lower Hutt, New Zealand). "The Design of Encoding Functions for Stereophonic and Polyphonic Sound Systems." November 1996.

Ville Pulkki (Laboratory of Acoustics and Audio Signal Processing, Helsinki University of Technology). "Virtual Sound Source Positioning Using Vector Base Amplitude Panning." June 1997.

K. Rasmussen and P. Juhl (Acoustics Laboratory, Technical University of Denmark). "The Effect of Head Shape on Spectral Stereo Theory." March 1993.

Gunther Theile (Institut fur Rundfunktechnik GmbH, Munich). "On the Naturalness of Two-Channel Stereo Sound." October 1991.

M. Wohr, G. Theile, H. Goeres, and A. Persterer. "Room-Related Balancing Technique: A Method for Optimizing Recording Quality." September 1991.

AES Conference on Spatial Sound Reproduction, April 10–12, 1999, Rovaniemi, Finland. Several papers.

Another excellent article is Geoff Martin's "Towards a Better Understanding of Stereo Microphone Technique," http://lecaine.music.mcgill.ca/~martin/my_stuff/mic_pairs.html.

A catalog of binaural recordings is available free from the The Binaural Source, Box 1727, Ross, CA 94957, tel. 415-457-9052.

13

SURROUND-SOUND TECHNIQUES, DVD, AND SUPER AUDIO CD

So far we examined two-channel recording techniques: stereo, binaural, and transaural. Stereo puts the instruments and hall reverb in front of the listener, in the area between the two loudspeakers. In contrast, binaural, transaural, and surround sound place audio images all around the listener. The musical ensemble usually is up front and the hall ambience is all around.

Stereo, binaural, and transaural use two channels feeding two loudspeakers. Surround sound uses multiple channels feeding multiple speakers.

A disadvantage of binaural is that you must wear headphones. A disadvantage of stereo and transaural is that you must sit in a tiny "sweet spot" to hear correct localization. In contrast, surround sound uses speakers and has a wide sweet spot.

Surround sound gives a wonderfully spacious effect. It puts you inside the concert hall with the musicians. You and the music occupy the same space—you're part of the performance. For this reason, surround sound is more musically involving, more emotionally intense, than regular stereo.

A magazine devoted to multichannel sound production is *Surround Professional* (6 Manhasset Ave., Port Washington, NY 11050, fax 516-767-1745).

Surround-Sound Speaker Arrangement

Inherited from the film industry, surround sound uses six channels feeding six speakers placed around the listener. This forms a 5.1 surround system, where "point 1" is the subwoofer or low-frequency effects (LFE) channel. The LFE channel is band limited to 120 Hz and below.

The six speakers are

- Left front.
- Center.
- Right front.
- Left surround.
- Right surround.
- Subwoofer.

Figure 13–1 shows the recommended placement of monitor speakers for 5.1 surround sound. It is the standard setup proposed by the International Telecommunication Union (ITU). From the center speaker, the left and right speakers should be placed at ±30°, and the surround speakers at ±110°.

The left-front and right-front speakers provide regular stereo. The surround speakers provide a sense of envelopment due to room ambience. They also allow sound images to appear behind the listener. Deep bass is filled in by the subwoofer. Because we do not localize low frequencies

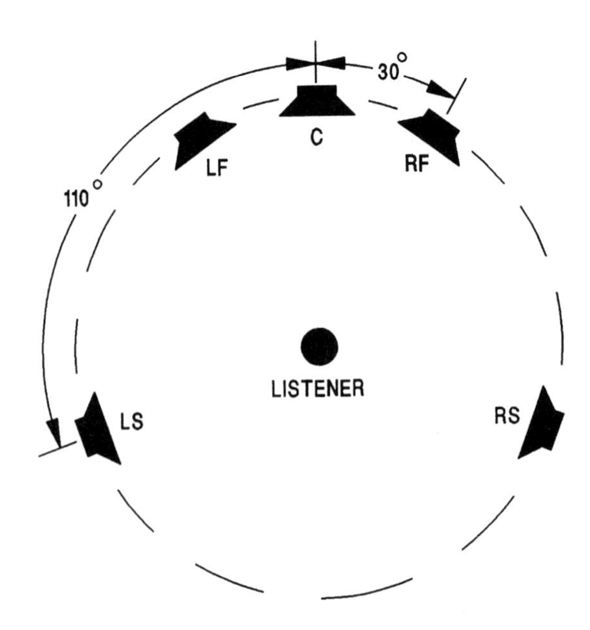

Figure 13–1
Recommended placement of monitor speakers for 5.1 surround sound.

below about 120 Hz, the subwoofer can be placed almost anywhere without degrading localization.

Originally developed for theaters, the center-channel speaker is mounted directly in front of the listener. In a home-theater system, it is placed just above or just below and in front of the TV screen. This speaker plays center-channel information in mono, such as dialogue.

Why use a center speaker when two stereo speakers create a phantom center image? If you use only two speakers and sit off center, the phantom image shifts toward the side on which you're sitting. But a center-channel speaker produces a real image, which does not shift as you move around the listening area. The center speaker keeps the actors' dialogue on screen, regardless of where the listener sits.

Also, the center speaker can have a flat response, but the phantom center image has a non-flat response. Why is this? Remember that a center image results when you feed identical signals to both stereo speakers. The right-speaker signal reaches your right ear, but so does the left-speaker signal after a delay around your head. The same thing occurs symmetrically at your left ear. Each ear receives direct and delayed signals, which interfere and cause phase cancellations at 2 kHz and above. A center-channel speaker does not have this response anomaly.

With a phantom center image, the response is weak at 2 kHz because of the phase cancellation just mentioned. To compensate, recording engineers often choose mics with a presence peak in the upper midrange for vocal recording. The center-channel speaker does not need this compensation.

For sharpest imaging and continuity of the sound field, all the speakers should be

- The same distance from the listener.
- The same model (except the subwoofer).
- The same polarity.
- Direct-radiator types.
- Driven by identical power amps.
- Matched in sound pressure level with pink noise.

A test tape for setting up multichannel speakers is available from TMH Corp. (213-742-0030, http://www.tmhlabs.com).

Typically the speakers are 4–8 feet from the listener and 4 feet high. Use a length of string to place the monitors the same distance from your head. The subwoofer can go along the front wall on the floor.

Be sure that all the speakers sound the same, so there is no change in tonal balance as you pan images around.

Surround-Sound Mic Techniques

In general, listening in surround sound reduces the stereo separation (stage width) because of the center speaker. Mic techniques for surround sound should be optimized to counteract this effect. A number of mic techniques have been developed for recording in surround sound. Let's take a look at them.

SoundField 5.1 Microphone System

This system is a single, multiple-capsule microphone (SoundField ST250 or MKV) and SoundField surround-sound decoder for recording in surround sound. The decoder translates the mic's B-format signals (X, Y, Z, and W) into L, C, R, LR, RR, and mono subwoofer outputs.

Delos VR² Surround-Sound Miking Method

John Eargle, Delos's director of recording, developed the firm's VR² (Virtual Reality Recording) format. Recordings made with this method offer discrete surround sound. They also are claimed to sound good in stereo and very good with "steered" analog decoding, such as Dolby Pro Logic (Long, 1996). In making these recordings, Eargle typically uses the mic placement shown in Figure 13–2.

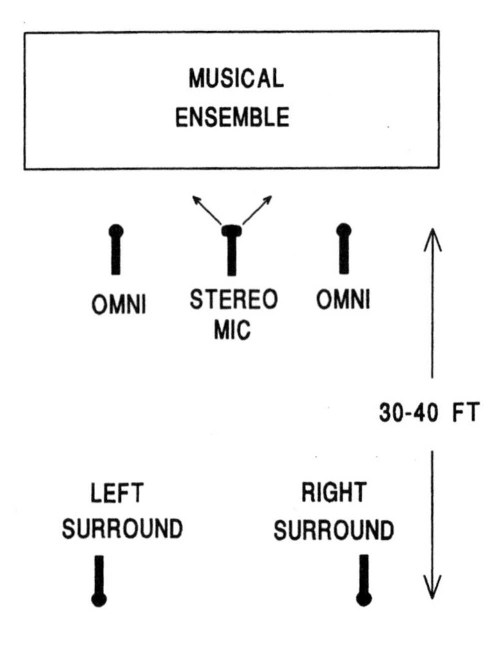

Figure 13–2
A Delos surround-sound miking method.

This method employs a coincident-stereo mic in the center, flanked by two spaced omnis, typically 4 feet apart. Two house mics (to pick up hall reverb) are placed 30–40 feet behind the main pair. (Greater spacing creates an undesirable echo.) These house mics are omnis or cardioids aiming at the rear of the hall, spaced about 12 feet apart. There also might be spot mics (accent mics) placed within the orchestra.

The mics are assigned to various tracks of a digital eight-track recorder:

1 and 2. A mix of the coincident-pair mics, flanking mics, house mics, and spot mics.

3 and 4. Coincident-pair stereo mic.

5 and 6. Flanking mics.

7 and 8. House mics (surround mics).

NHK Method

The Japanese NHK Broadcast Center has worked out another surround-sound miking method (Fukada, Tsujimoto, and Akita, 1997). They found, for surround-sound recording, that cardioid mics record a more natural amount of reverb than omni mics (Fukada, 1997). The mics are placed as follows (see Figure 13–3):

- A center-aiming mic feeds the center channel.
- A near-coincident pair feeds front-left and front-right channels.
- Widely spaced flanking mics add expansiveness.
- Up to three pairs of ambience mics aim toward the rear.

The front mics are placed at the critical distance from the orchestra, where the direct-sound level matches the reverberant-sound level. Typically, this point is 12–15 feet from the front of the musical ensemble and 15 feet above the floor.

NHK engineers recommend that, when you're monitoring the surround-sound program, the reverb volume in stereo listening should match the reverb volume in multichannel listening. That is, when you fold down or collapse the monitoring from 5.1 to stereo, the direct/reverb ratio should stay the same.

The KFM 360 Surround-Sound System

Jerry Bruck (1997) of Posthorn Recordings developed this elegant surround-sound miking method. It is a form of the mid-side (MS) stereo technique.

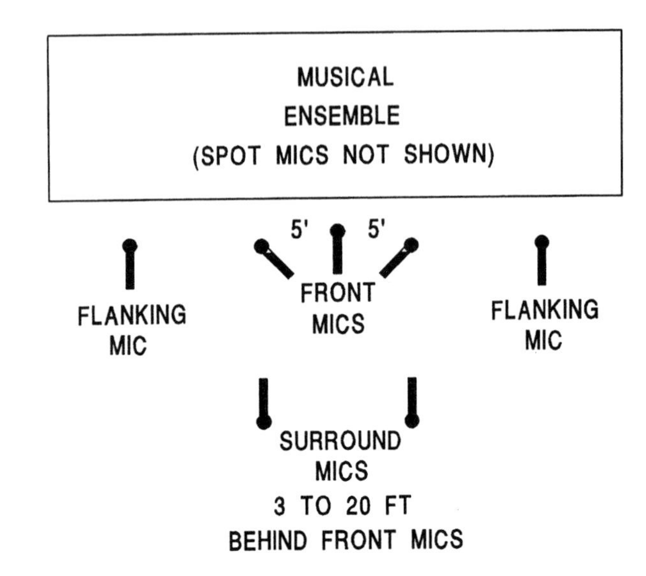

Figure 13–3
An NHK surround-sound miking method.

Bruck starts with a modified Schoeps KFM 6U stereo microphone, which is a pair of omni mics mounted in opposite sides of a 7 inch hard sphere. Next to those mics, nearly touching, are two figure-eight mics, one on each side of the sphere, each aiming front and back (Figure 13–4). This array creates two MS mic arrays aimed sideways in opposite directions. The mics do not seriously degrade each other's frequency response.

In the left channel, the omni and figure-eight mic signals are summed to give a front-facing cardioid pattern. They also are differenced to give a rear-facing cardioid pattern. The same thing happens symmetrically in the right channel.

The sphere, acting as a boundary and a baffle, "steers" the cardioid patterns off to either side of center and makes their polar patterns irregular.

By adjusting the relative levels of the front and back signals, the user can control the front/back separation. As a result, the mic sounds like it is moving closer to or farther from the musical ensemble.

According to Bruck,

> The system is revelatory in its ability to recreate a live event. Perhaps most remarkable is the freedom a listener has to move around and select a favored position, as one might move around in a concert hall to select a preferred seat. The image remains stable, without a discernible "sweet spot." The reproduction is unobtrusively natural and convincing in its sense of "being there."

The four mic signals can be recorded on a four-track recorder for later matrixing.

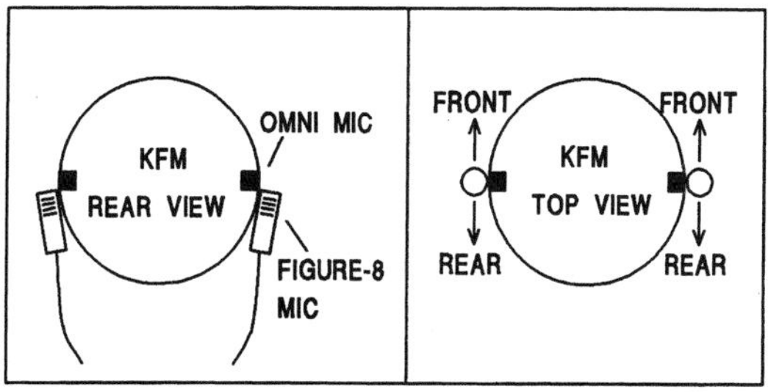

Figure 13–4
KFM 360 surround-sound miking method.

The figure-eight mics need some equalization to compensate for their low-frequency rolloff and loss in the extreme highs. To maintain a good signal-to-noise ratio, this EQ can be applied after the signal is digitized.

Five-Channel Microphone Array with a Binaural Head

This method, developed by John Klepko (1997) of McGill University, combines an array of three directional mics with a two-channel dummy head (Figure 13–5), where

The front left and right channels are identical supercardioid mics.

The center channel is a cardioid mic.

The surround channels comprise a dummy head with two pressure-type omni mics fitted into the ear molds.

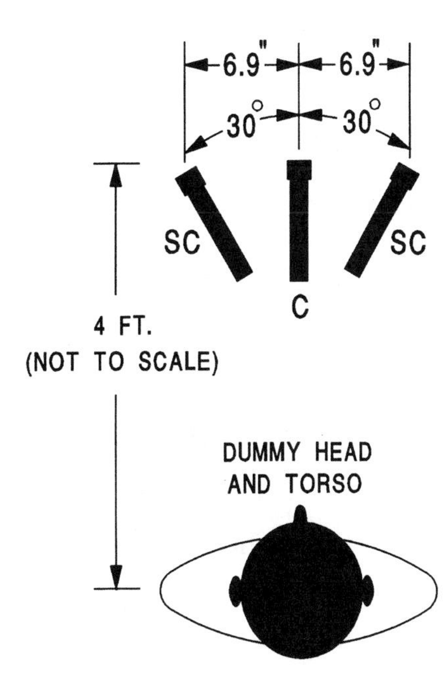

Figure 13–5
The Klepko surround-sound miking method.

The mics are shock mounted and have equal sensitivity and equal gains. Supercardioids are used for the front left and right pair to reduce center-channel buildup.

Although the dummy head's diffraction causes peaks and dips in the response, it can be equalized to compensate for them.

During playback, the listener's head reduces the acoustical crosstalk that normally would occur between the surround speakers.

According to Klepko,

> The walkaround tests form an image of a complete circle of points surrounding the listening position. Of particular interest is the imaging between ±30° and ±90°. The array produces continuous, clear images here where other [surround-sound] techniques fail.
>
> The proposed approach is downward compatible to stereo although there will be no surround effect. However, stereo headphone reproduction will resolve a full surround effect due to the included binaural head-related signals. Downsizing to matrix multichannel (5-2-4 in this case) is feasible except that it will not properly reproduce binaural signals to the rear because of the mono surrounds. As well, some of the spatial detail recorded by the dummy-head microphone would be lost due to the usual bandpass filtering scheme (100 Hz–7 kHz) of the surround channel in such matrix systems.

DMP Method

DMP engineer Tom Jung (1997) has recorded in surround sound using a Decca tree stereo array for the band and a rear-aiming stereo pair for the surround ambience (Figure 13–6). Spot mics in the band complete the miking. The Decca tree uses three mics spaced a few feet apart, with the center mic placed slightly closer to the performers. It feeds the center channel in the 5.1 system.

The rear-aiming mic pair is either a coincident stereo mic or a spaced pair whose spacing matches that of the Decca tree outer pair. Jung tries to aim the rear mics at irregular surfaces to pick up diffuse sound reflections.

Woszcyk Technique (PZM Wedge Plus Opposite-Polarity, 180° Coincident-Cardioid Surround-Sound Mics)

A recording instructor at McGill University, Wieslaw Woszcyk (1990) developed an effective method for recording in surround sound that also

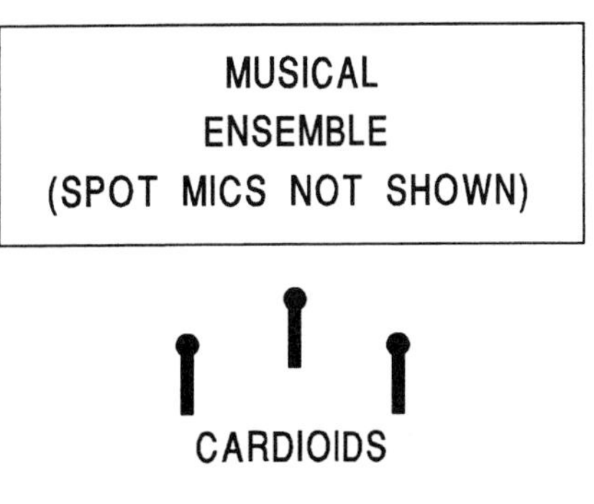

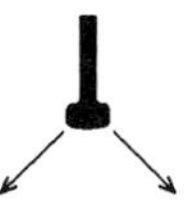

Figure 13–6
A DMP surround-sound miking method.

works well in stereo. The orchestra is picked up by a PZM wedge made of two 18 x 29 inch hard baffle boards angled 45°. A mini omni mic is mounted on or flush with each board. At least 20 feet behind the wedge are the surround-sound mics: two coincident cardioids angled 180° apart, aiming left and right, and in opposite polarity (Figure 13–7).

According to Woszcyk, his method has several advantages:

- Imaging is very sharp and accurate, and spaciousness is excellent due to strong pickup of lateral reflections.
- The out-of-phase impression of the surround-sound pair disappears when a center coherent signal is added.
- The system is compatible in surround sound, stereo, and mono. In other words, the surround-sound signals do not phase interfere with the front-pair signals. That is because (1) the surround-sound signals are delayed more than 20 milliseconds, (2) the two mic pairs operate in separate sound fields, and (3) the surround-sound mics form a bidirectional pattern in mono, with its null aiming at the sound source.

If a PZM wedge is not acceptable because of its size and weight, other arrays with wide stereo separation may be substituted.

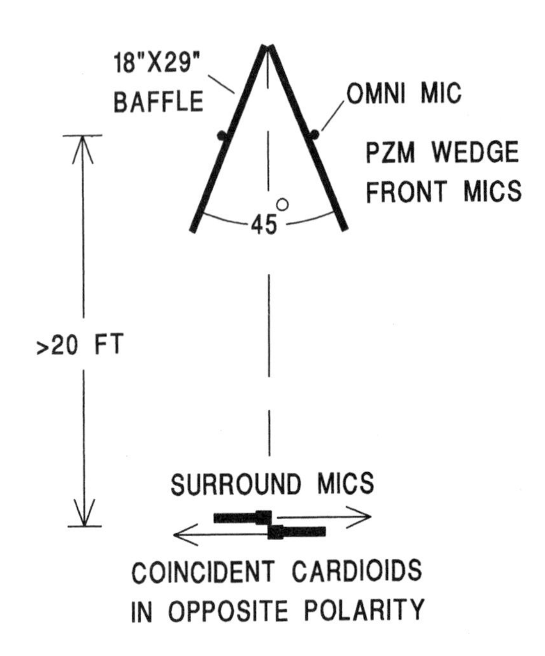

Figure 13–7
Woszcyk surround-sound miking method.

So far we discussed surround recordings and some mic techniques used to make them. Now we turn our attention to the media that will play those surround recordings.

Surround-Sound Media

Surround-sound recordings can be distributed to the public on super-audio compact disc or Digital Versatile Disc (DVD, explained later). For surround programs, both media use 5.1-channel encoding:

1. Left front.
2. Right front.
3. Center.
4. Left surround.
5. Right surround.

6 or 1. Subwoofer or LFE channel.

The encoding format used in DVD is Dolby Digital (AC-3), DSD (Direct Stream Digital), or digital theater systems (DTS). Each channel in the 5.1 scheme is digital and discrete, not matrixed.

In contrast, Dolby Surround is a matrix encoding system that combines four channels (left, center, right, surround) into two channels. The

Dolby Pro Logic decoder unfolds the two channels back into four. The surround channel, which has limited bandwidth, is reproduced over left and right surround speakers.

First used in movie theaters, Dolby Digital and DTS are perceptual coding methods. They use data compression to remove sounds deemed inaudible due to masking. DTS compresses the data with about a 3:1 ratio. Both formats offer six discrete channels of digital surround sound. DTS resolution is 20 bits, while Dolby Digital is 16, 18, or 20 bits (perhaps 24 bits in the future). Dolby's encoder accepts data at sample rates of 32, 44.1, or 48 kHz; DTS accepts 44.1 kHz.

To play a Dolby Digital or DTS recording, you take the digital output of a DVD player or CD player into a decoder. The decoder has a DSP chip that can decode both the DTS and Dolby Digital formats. Some surround receivers have DTS and Dolby Digital decoders built in.

The Digital Versatile Disc

A medium for playing surround recordings is the Digital Versatile Disc, or DVD. It is a high-capacity optical storage medium the size of a compact disc (4.73 inches diameter). DVD can store digital data in three formats: audio, video, and computer.

DVD Compatibility

The DVD player reads CD, CD-R (in some units), CD-ROM, DVD-video disk (video plus audio), DVD-audio disc (audio only).

DVD Capacity

DVD has much greater capacity than a CD, due to its smaller pits and closer tracks. The scanning laser has a shorter wavelength than in a CD player, which lets it read DVD's denser data stream. In addition, MPEG-2 data compression of the video data increases the data densitity on the DVD.

Some DVDs have a single layer of pits; others have a dual layer at different depths. The laser automatically focuses on the required layer.

The storage capacities of various media follow:

- A compact disc holds 640 MB.
- A single-sided, single-layer DVD holds 4.7 GB (seven times CD capacity).
- A single-sided, dual-layer DVD holds 8.5 GB.

- A double-sided, single-layer DVD holds 9.4 GB.
- A double-sided, dual-layer DVD holds 17 GB.

The single-sided, single-layer disc was the main DVD format when this book went to print. Its 4.7 GB capacity is enough to hold a 2 hour, 10 minute movie plus subtitles.

DVD Players

Most DVD players have these types of output:

- Two-channel analog output playing a Dolby Surround-encoded stereo signal.
- A digital output that can be connected to an external decoder for 5.1-channel surround sound (either Dolby Digital or DTS).
- Six analog channels fed from built-in Dolby Digital or DTS decoders (in some units).

DVD-Audio Discs

DVD Audio is an audio-only DVD format using linear PCM encoding. A DVD-audio specification (version 1.0) was approved in Fall 1998. It uses a lossless data-compression scheme called Meridian Lossless Packing. Compared to standard CD audio, DVD-audio permits

- Better fidelity due to higher sampling rate and higher quantization.
- Longer playing time in some formats.
- More channels (up to six in the 5.1 scheme).
- Optional text and still pictures.
- Optional formats such as DSD (Direct Stream Digital), MPEG-2 BC, lossless compression, DTS, and Dolby Digital.

DVD audio allows a wide range of sampling rates, number of channels, and quantizations. For example, a 4.7 GB single-layer disc can hold a 75 minute program in which the left-center-right signals are 88.2 kHz/24 bit and the two surround channels are 44.1 kHz/20 bit.

Audio on DVD-Video Discs

On a DVD-video disc, you can have either Dolby Digital (AC-3) surround-sound, 5.1 channels, with MPEG-1 compressed audio, or two channels, 16 to 24 bit, 48 or 96 kHz, linear PCM audio. In other words, you can have surround sound with compromised fidelity or two-channel sound with excellent fidelity. A DVD disk can be encoded with both formats, and the listener can choose which one to hear.

Although the spec provides for multichannel PCM audio, current players have only two channels of PCM audio. All players support Dolby Digital (AC-3) surround sound.

Data Rate

The DVD data transfer rate is 10.1 megabits per second (mps). A six-channel surround-sound track, with data compression, consumes 384 or 448 kilobits per second (kps). Linear PCM stereo, at 16 bits and 48 kHz sampling rate, takes 1.5 mps. A 96 kHz sample rate doubles the transfer rate to 3 mps. A 24-bit resolution increases the rate to 4.5 mps.

DVD Premastering Formats

These are acceptable media formats to send to a DVD manufacturer. For two-channel stereo, use one of these formats:

- A removable hard drive with SMPTE time code.
- DAT with SMPTE.
- The digital audio tracks on the video master tape (D-1, D-5, digital Beta-Cam, DCT or D-2).

For 5.1-channel surround, you can use one of these formats:

- 20-bit Alesis ADAT with time code.
- Genex magneto-optical eight-track 20-bit machine.
- DLT, DVD-R, or Sonic Solutions 24- or 20-bit Exabyte.
- TASCAM DA-88 (or equivalent). Record the six audio channels on tracks 1 through 6. Put SMPTE on track 8.
- Or use all eight DA-88 tracks to simulate six tracks with 20-bit resolution. This is done with a Prism MR-2024T processor. For Dolby Digital, track assignments are 1 = left, 2 = center, 3 = right, 4 = LFE (low-frequency effects or subwoofer), 5 = left surround, 6 = right surround, 7 = data channel, 8 = data channel. For DTS, the track assignment is 1 = left, 2 = right, 3 = left surround, 4 = right surround, 5 = center, 6 = LFE, 7 = data, 8 = data.
- Pro Tools 24 with Jaz drives.
- Nagra-D is fine for four-channel surround sound.

To create the LFE or .1 channel, you can mix the five other channels and feed them through a 120 Hz low-pass filter.

You might have video slide-show data that accompanies the audio. Store the video on a removable hard drive or send the file by modem on the Internet. TIFF and bitmap (BMP) formats are acceptable.

Sonic Solutions makes Sonic Studio software editors for DSD and DVD-Audio authoring. They handle several formats including DSD, four channels of high-densitity audio (24-bit, 96 kHz), Dolby Digital 5.1 surround sound, MPEG-2 stereo, and PCM (44.1, 48, or 96 kHz). Using a graphical interface, Sonic Studio provides editing and processing such as level balancing, surround panning, No-Noise, and pitch shift. Sonic Solutions is at 101 Rowland Way, Novato, CA 94945, tel. 415-893-8000, http://www.sonic.com.

Dolby Units for DVD Mastering

DVD mastering requires MPEG-2 video and Dolby Digital encoders. You can use either one for multichannel audio on DVD titles. Dolby Digital soundtracks are compatible with mono, stereo, or Dolby Pro Logic systems.

Some Dolby encoders and decoders include

DP561B: A free-standing 5.1-channel AC-3 audio encoder with data compression.

DP569: A 5.1-channel Dolby Digital encoder.

DP567: A two-channel audio encoder.

DP562: A multichannel Dolby Digital and Dolby Surround decoder. Up to 5.1-channels decoding (for monitoring) and four-channel Dolby Pro Logic decoding.

SEU4 and SDU4 Dolby Surround encoder-decoder units.

Dolby Surround tools TDM plug-ins for Digidesign Pro Tools workstations.

Super Audio CD

An alternative to DVD audio is the Super Audio CD proposed by Sony and Philips. It uses the direct stream digital (DSD) process, which encodes a digital signal in a 1-bit (bitstream) format at a 2.8224 MHz sampling rate. This system offers a frequency response from DC to 100 kHz with a 120 dB dynamic range.

The Super Audio CD has two layers that are read from one side of the disc. On one layer is a two-channel stereo DSD program followed by the same program in six channels for surround sound. On the other layer is a Red-Book 16 bit/44.1 K program, which makes dual inventories unnecessary. The 16/44.1 K signal is derived from Sony's Super Bit Mapping Direct processing, which downconverts the bitstream signal with minimal loss of DSD quality. Standard CD players can play the 16/44.1 K layer, and future players would be able to play the two- or six-channel DSD layer for even better sound quality.

References

Jerry Bruck. "The KFM 360 Surround—A Purist Approach." Preprint 4637 (F-3) from the 103rd convention of the Audio Engineering Society, September 1997.

Akira Fukada. "Our Challenges for Multichannel Music Mixing and the Subject of Expression." Paper presented at the Audio Engineering Society Tokyo convention, June 1997.

Akira Fukada, Kiyoshi Tsujimoto, and Shoji Akita (NHK Broadcast Center). "Microphone Techniques for Ambient Sound on a Music Recording." Preprint 4540 (F-2) from the 103rd convention of the Audio Engineering Society, September 1997.

Tom Jung. "Mixing with the Big Bands." *EQ* (October 1997), pp. 84–88.

John Klepko. "Five-Channel Microphone Array with Binaural Head for Multichannel Reproduction." Preprint 4541 (F-4) from the 103rd convention of the Audio Engineering Society, September 1997.

Ed Long. "Rushing to Virtual Reality." *Audio* (June 1996), p. 32.

Wieslaw Woszcyk. "A New Method for Spatial Enhancement in Stereo and Surround Recording." Preprint 2946, paper presented at the Audio Engineering Society 89th convention, September 21–25, 1990, Los Angeles.

The information on surround sound was taken from the following sources:

Ed Cherney. "Surround Gets Road Tested." *EQ* (October 1997), pp. 90–94.

Steve La Cerra. "Get a Feel for Surround." *EQ* (Oct. 1997), p. 8.

Robert Margouleff and Brant Biles with Steve La Cerra. "Creating Surround Sound Mixes." *EQ* (October 1997), pp. 72–78.

Doug Mitchell. "Teaching Concepts in Mixing for Surround Sound—Pedagogical Changes with the Shift to 5.1." Preprint 4544 (F-10) from the Audio Engineering Society103rd convention, September 1997.

Surround Professional no. 1 (October 1998).

David Tickle. "The Joy of Six." *EQ* (October 1997), pp. 96–107.

Frank Wells. "Surround Sound Comes of Age." *Audio Media* (October 1997), pp. 128–135.

The information on DVD came from these sources:

Philip De Lancie. "DVD-Audio, Format in Search of an Identity." *Mix* (February 1998), pp. 44–212.

Philip De Lancie. "Meridian Lossless Packing." *Mix* (December 1998), pp. 80-88.

Digital Theater Systems. "Digital DTS Surround." Brochure from www.dtstech.com.

Dolby Laboratories. *Dolby News* (Fall 1997).

Gary Hall."Understanding DVD." *Mix* (September 1997), pp. 54-225.

Oliver Masciarotte. "DVD—Closing in on a Final Spec." *Mix* (July 1998), pp. 130–137.

Ian G. Masters. "Digital Surrounds" (part of "Audio Q&A"). *Stereo Review* (April 1998), p. 28.

The information on Super Audio CD and DSD was taken from the white paper "Super Audio Compact Disc: A Technical Proposal" by Sony/Philips, 1997.

14

STEREO RECORDING PROCEDURES

This chapter is divided into three parts. The first part explains how to do an on-location stereo recording of a classical music ensemble (Bartlett, 1998). The second part covers the basics of stereo miking for popular music. The last part is a special troubleshooting guide to help you pinpoint and solve problems in stereo reproduction.

We start by going over the equipment and procedures for an on-location recording of classical music.

Equipment

Before going on location, you need to assemble a set of equipment, such as this:

- Microphones (low-noise condenser or ribbon type, omni or directional, free field or boundary, stereo or separate).
- Recorder (open reel, DAT, MDM, etc., noise reduction optional).
- Low-noise mic preamps (unless the mic preamps in your recorder are very good).
- Phantom power supply (unless your mic preamp or mixer has phantom power built in).
- Mic stands and booms or fishing line, stereo bar, shock mounts (optional).

- Microphone extension cables.
- Mixer (optional).
- MS matrix box (optional).
- Headphones or speakers.
- Power amplifier for speakers (optional).
- Blank tape.
- Stereo phase-monitor oscilloscope (optional).
- Power strip, extension cords.
- Notebook and pen.
- Tool kit.

First on the list are microphones. You need at least two or three of the same model number or one or two stereo microphones. Good microphones are essential, for the microphones—and their placement—determine the sound of your recording. You should expect to spend at least $250 per microphone for professional-quality sound.

For classical music recording, the preferred microphones are condenser or ribbon types with a wide, flat frequency response and very low self-noise (explained in Chapter 7). A self-noise spec of less than 21 dB equivalent SPL, A-weighted, is recommended.

You need a power supply for condenser microphones: either an external phantom power supply, a mixer or mic preamp with phantom power, or internal batteries.

If you want to do spaced-pair recording, you can use either omnidirectional or directional microphones. Omnis are preferred because they generally have a flatter low-frequency response. If you want to do coincident or near-coincident recording for sharper imaging, use directional microphones (cardioid, supercardioid, hypercardioid, or bidirectional).

You can mount the microphones on stands or hang them from the ceiling with nylon fishing line. Stands are much easier to set up but more visually distracting at live concerts. Stands are more suitable for recording rehearsals or sessions with no audience.

The mic stands should have a tripod folding base and extend at least 14 feet high. To extend the height of regular mic stands, you can either use baby booms or telescoping photographic stands (available from camera stores). These are lightweight and compact.

A useful accessory is a stereo bar or stereo microphone adapter. This device mounts two microphones on a single stand for stereo recording. Another needed accessory in most cases is a shock mount to prevent pick-up of floor vibrations.

In difficult mounting situations, boundary microphones may come in handy. They can lie flat on the stage floor to pick up small ensembles or be mounted on the ceiling or on the front edge of a balcony. They also can be attached to clear plexiglass panels hung or mounted on mic stands.

For monitoring in the same room as the musicians, you need some closed-cup, circumaural (around the ear) headphones to block out the sound of the musicians. You want to hear only what is being recorded. Of course, the headphones should be wide range and smooth for accurate monitoring. A better monitoring arrangement might be to set up an amplifier and close-field loudspeakers in a separate room.

If in the same room as the musicians, you have to sit far from the musicians to clearly monitor what you're recording. To do that, you need a pair of 50 foot microphone extension cables. Longer extensions are needed if the mics are hung from the ceiling or if you're monitoring in a separate room.

A mixer is necessary when you want to record more than one source; for example, an orchestra and a choir or a band and a soloist. You might put a pair of microphones on the orchestra and another pair on the choir. The mixer blends the signals of all four mics into a composite stereo signal. It also lets you control the balance (relative loudness) among microphones. A mixer is also needed if you plan to do spot miking.

For monitoring a mid-side recording, bring an MS matrix box that converts the MS signals to L/R signals, which you monitor.

Note: Be sure to test all your equipment for correct operation before going on the job.

Choosing the Recording Site

If possible, plan to record in a venue with good acoustics. There should be adequate reverberation time for the music being performed (at least 2 seconds for orchestral recording). This is very important, because it can make the difference between an amateur-sounding recording and a commercial-sounding one. Try to record in an auditorium, concert hall, or spacious church rather than in a band room or gymnasium. Avoid stage shells because they lack a sense of space.

You may be forced to record in a hall that is too dead; that is, the reverberation time is too short. In this case, you may want to add artificial reverberation from a digitial reverb unit or cover the seats with plywood sheets or 4 millimeter polyethylene plastic sheeting. Strong echoes can be controlled with carpets, RPG diffusors, or drapes. Dry climates tends to shorten the reverb time and dull the sound.

Session Setup

If the orchestral sound from the stage is bad, you might want to move the orchestra out onto the floor of the hall.

Place your microphones in the desired stereo miking arrangement. As an example, suppose you're recording an orchestra rehearsal with two crossed cardioids on a stereo bar (the near-coincident method). Screw the stereo bar onto a mic stand and mount two cardioid microphones on the stereo bar. For starters, angle them 110° apart and space them 7 inches apart horizontally. Aim them downward so that they'll point at the orchestra when raised. You may want to mount the microphones in shock mounts or put the stands on sponges to isolate the mics from floor vibration.

Basically, you want to place two or three mics several feet in front of the group, raised up high (as in Figure 14–1).The microphone placement

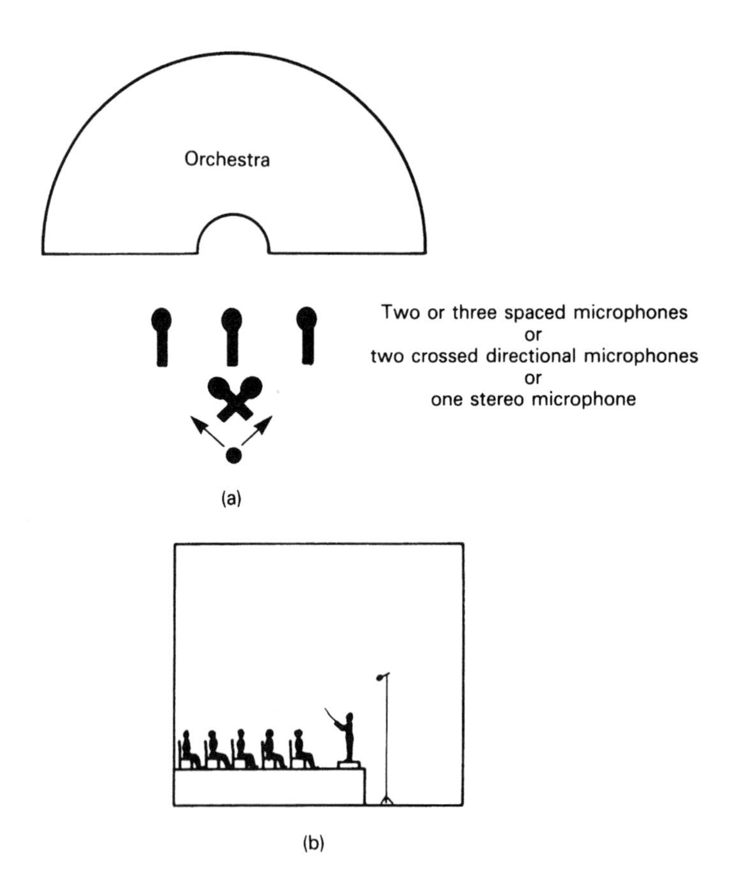

Figure 14–1 Typical microphone placement for on-location recording of a classical music ensemble: (a) top view, (b) side view.

controls the acoustic perspective or sense of distance to the ensemble, the balance among instruments, and the stereo imaging.

As a starting position, place the mic stand behind the conductor's podium, about 12 feet from the front-row musicians. Connect mic cables and mic extension cords. Raise the microphones about 14 feet off the floor. This prevents overly loud pickup of the front row relative to the back row of the orchestra.

Leave some extra turns of mic cable at the base of each stand so you can reposition the stands. This slack also allows for people accidentally pulling on the cables. Try to route the mic cables where they won't be stepped on, or cover them with mats.

Live, broadcast, or filmed concerts require an inconspicuous mic placement, which may not be sonically ideal. In these cases, or for permanent installations, you probably want to hang the microphones from the ceiling rather than using stands. You can hang the mics by their cables or nylon fishing line of sufficient tensile strength to support their weight. Another inconspicuous placement is on mic-stand booms projecting forward from a balcony in front of the stage. For dramas or musicals, directional boundary mics can be placed on the stage floor near the footlights.

Now you're ready to make connections. There are several different ways to do this:

- If you're using just two mics, you can plug them directly into a phantom supply (if necessary) and from there into your tape deck. You might prefer to use low-noise mic preamps, then connect cables from there into your recorder line inputs.
- If you're using two mics and a noise-reduction unit, plug the mics into a mixer or preamps to boost the mic signals up to line level, then run that line-level signal into the noise-reduction unit connected to the recorder line inputs.
- If you're using multiple mics (either spot mics or two MS mics) and a mixer, plug the mics into a snake box (described in Chapters 1 and 7). Plug the mic connectors at the other end of the snake into your mixer mic inputs. Finally, plug the mixer outputs into the recorder line inputs.
- If you're also using noise reduction, plug the mixer outputs into the inputs of the noise-reduction device and from there into the recorder.
- If you want to feed your mic signals to several mixers—for example, one for recording, one for broadcast, and one for sound reinforcement—plug your mic cables into a mic splitter or distribution amp (described in Chapter 2). Connect the splitter output to the snakes for each mixer. Supply phantom power from one mixer only, on the

microphone side of the split. Each split will have a ground-lift switch on the splitter. Set it to *ground* for only one mixer (usually the recording mixer). Set it to *lift* for the other mixers. This prevents hum caused by ground loops between the different mixers.

- If you're using directional microphones and want to make their response flat at low frequencies, you can run them through a mixer with equalization for bass boost. Boost the extreme low frequencies until the bass sounds natural or until it matches the bass response of omni condenser mics. Connect the mixer output either into an optional noise-reduction unit or directly into your recorder. This equalization will be unnecessary if the microphones have been pre-equalized by the manufacturer for flat response at a distance.

Monitoring

Put on your headphones or listen over loudspeakers in a separate room. Sit equidistant from the speakers, as far from them as they are spaced apart. You probably need to use a close-field arrangement (speakers about 3 feet apart and 3 feet from you) to reduce coloration of the speakers' sound from the room acoustics.

Turn up the recording-level controls and monitor the signal. When the orchestra starts to play, set the recording levels to peak roughly around –10 VU so you have a clean signal to monitor. You can set levels more carefully later on.

Microphone Placement

Nothing has more effect on the production style of a classical music recording than microphone placement. Miking distance, polar patterns, angling, spacing, and spot miking all influence the recorded sound character.

Miking Distance

The microphones must be placed closer to the musicians than a good live listening position. If you place the mics out in the audience where the live sound is good, the recording probably will sound muddy and distant when played over speakers. That is because all the recorded reverberation is reproduced up front along a line between the playback speakers, along with the direct sound of the orchestra. Close miking (5–20 feet from the front row) compensates for this effect by increasing the ratio of direct sound to reverberant sound.

The closer the mics are to the orchestra, the closer it sounds in the recording. If the instruments sound too close, too edgy, or too detailed, or if the recording lacks hall ambience, the mics are too close to the ensemble. Move the mic stand a foot or two farther from the orchestra and listen again.

If the orchestra sounds too distant, muddy, or reverberant, the mics are too far from the ensemble. Move the mic stand a little closer to the musicians and listen again.

Eventually you'll find a sweet spot, where the direct sound of the orchestra is in a pleasing balance with the ambience of the concert hall. Then the reproduced orchestra will sound neither too close nor too far.

Here's why miking distance affects the perceived closeness (perspective) of the musical ensemble: The level of reverberation is fairly constant throughout a room, but the level of the direct sound from the ensemble increases as you get closer to it. Close miking picks up a high ratio of direct-to-reverberant sound; distant miking picks up a low ratio. The higher the direct-to-reverb ratio, the closer the sound source is perceived to be.

An alternative to finding the sweet spot is to place a stereo pair close to the ensemble (for clarity) and another stereo pair distant from the ensemble (for ambience). According to Delos recording director John Eargle, the distant pair should be no more than 30 feet from the main pair. You mix the two pairs with a mixer. The advantages of this method are

- It avoids pickup of bad-sounding early reflections.
- It allows remote control (via mixer faders) of the direct/reverb ratio or the perceived distance to the ensemble.
- Comb filtering due to phase cancellations between the two pairs is not severe because the delay between them is great and their levels and spectra are different. If the distant pair is farther back than 30 feet, its signal might simulate an echo.

Similarly, Skip Pizzi (1984) recommends a "double MS" technique, which uses a close MS microphone mixed with a distant MS microphone (as shown in Figure 14–2). One MS microphone is close to the performing ensemble for clarity and sharp imaging, and the other is out in the hall for ambience and depth. The distant mic could be replaced by an XY pair for lower cost.

If the ensemble is being amplified through a sound-reinforcement system, you might be forced to mike very close to avoid picking up amplified sound and feedback from the reinforcement speakers.

For broadcast or communications, consider miking the conductor with a wireless lavalier mic.

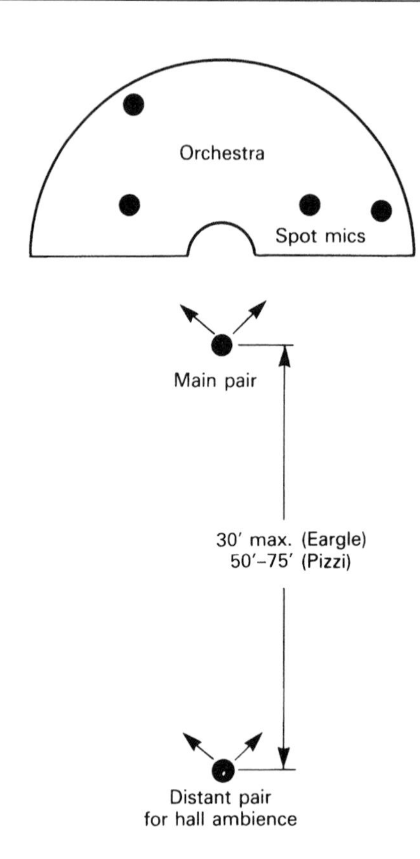

Figure 14–2
Double MS technique using a close main pair and a distant pair for ambience. Spot mics also are shown.

Stereo-Spread Control

Concentrate on the stereo spread. If the monitored spread is too narrow, it means that the mics are angled or spaced too close together. Increase the angle or spacing between mics until localization is accurate. Note: Increasing the angle between mics will make the instruments sound farther away; increasing the spacing will not.

If off-center instruments are heard far left or far right, that indicates your mics are angled or spaced too far apart. Move them closer together until localization is accurate.

If you record with a mid-side microphone, you can adjust the stereo spread by remote control at the matrix box with the stereo spread control (M/S ratio control).

You can change the monitored stereo spread either during the recording or after:

- To change the spread during the recording, connect the stereo-mic output to the matrix box and connect the matrix-box output to the recorder. Use the stereo-spread control (M/S ratio) in the matrix box to adjust the stereo spread.

- To alter the spread after the recording, record the middle signal on one track and the side signal on another track. Monitor the output of the recorder with a matrix box. After the recording, run the middle and side tracks through the matrix box, adjust the stereo spread as desired, and record the result.

If you are set up before the musicians arrive, check the localization by recording yourself speaking from various positions in front of the microphone pair while announcing your position (e.g., "left side," "mid left," "center"). Play back the recording to judge the localization accuracy provided by your chosen stereo array. Recording this localization test at the head of a tape is an excellent practice.

Monitoring Stereo Spread

Full stereo spread on speakers is a spread of images all the way between speakers, from the left speaker to the right speaker. Full stereo spread on headphones can be described as stereo spread from ear to ear. The stereo spread heard on headphones may or may not match the stereo spread heard over speakers, depending on the microphone technique used.

Due to psychoacoustic phenomena, coincident-pair recordings have less stereo spread over headphones than over loudspeakers. Take this into account when monitoring with headphones or use only loudspeakers for monitoring.

If you are monitoring the recording over headphones or anticipate headphone listening to the playback, you may want to use near-coincident miking techniques, which have similar stereo spread on headphones and loudspeakers.

Ideally, monitor speakers should be set up in a close-field arrangement (say, 3 feet from you and 3 feet apart) to reduce the influence of room acoustics and to improve stereo imaging.

If you want to use large monitor speakers placed farther away, deaden the control-room acoustics with Sonex or thick fiberglass insulation (covered with muslin). Place the acoustic treatment on the walls behind and to the sides of the loudspeakers. This smoothes the frequency response and sharpens stereo imaging.

You'll probably want to include a stereo-mono switch in your monitoring system, as well as an oscilloscope. The oscilloscope is used to check for excessive phase shift between channels, which can degrade mono frequency response or cause record-cutting problems. Connect the left-channel signal to the oscilloscope's vertical input; connect the right-channel signal to the horizontal input, and look for the Lissajous patterns shown in Figure 14–3.

Soloist Pickup and Spot Microphones

Sometimes a soloist plays in front of the orchestra. You have to capture a tasteful balance between the soloist and the ensemble. That is, the main stereo pair should be placed so that the relative loudness of the soloist and the accompaniment is musically appropriate. If the soloist is too loud relative to the orchestra (as heard on headphones or loudspeakers), raise the

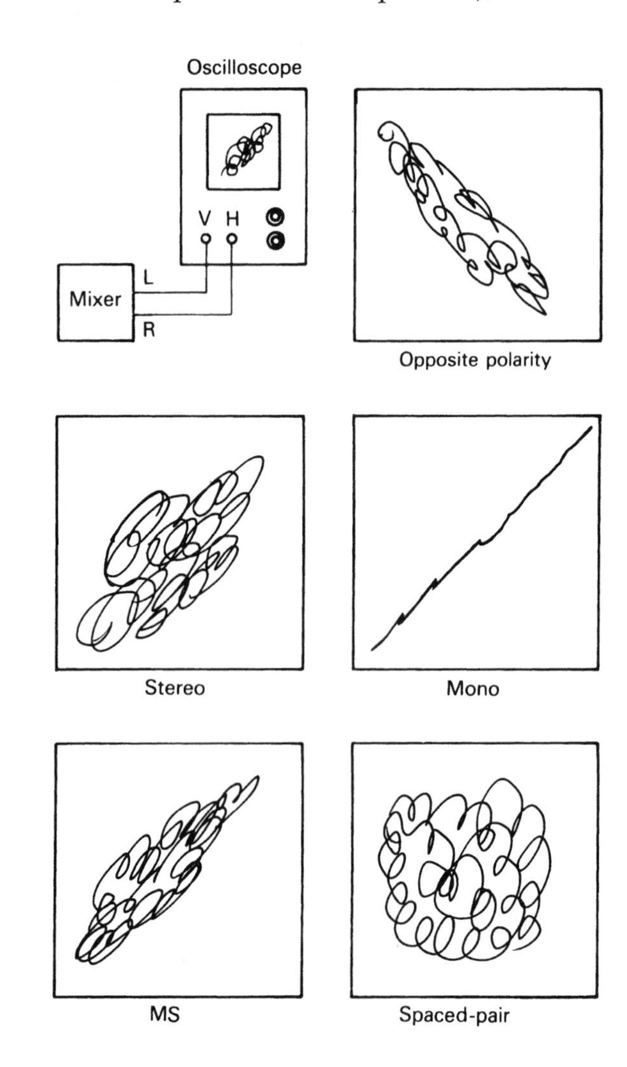

Figure 14–3
Oscilloscope Lissajous patterns showing various phase relationships between channels of a stereo program.

mics. If the soloist is too quiet, lower the mics. You may want to add a spot mic (accent mic) about 3 feet from the soloist and mix it with the other microphones. Take care that the soloist appears at the proper depth relative to the orchestra.

Many record companies prefer to use multiple microphones and multitrack techniques when recording classical music. Such methods provide extra control of balance and definition and are necessary in difficult situations. Often, you must add spot or accent mics on various instruments or instrumental sections to improve the balance or enhance clarity (as shown in Figure 14–2). In fact, John Eargle contends that a single stereo pair of mics rarely works well.

A choir that sings with an orchestra can be placed behind the orchestra; in this case, mike with two to four cardioids. Or the choir can stand in the audience area facing the orchestra.

Pan each spot mic so that its image position coincides with that of the main microphone pair. Using the mute switches on your mixing console, alternately monitor the main pair and each spot to compare image positions.

You might want to use an MS microphone or stereo pair for each spot mic and adjust the stereo spread of each local sound source to match that reproduced by the main pair. For example, suppose that a violin section appears 20° wide as picked up by the main pair. Adjust the perceived stereo spread of the MS spot mic used on the violin section to 20°, then pan the center of the section image to the same position that it appears with the main mic pair.

When you use spot mics, mix them at a low level relative to the main pair—just loud enough to add definition but not loud enough to destroy depth. Operate the spot-mic faders subtly or leave them untouched. Otherwise, the close-miked instruments may seem to jump forward when the fader is brought up, then fall back in when the fader is brought down. If you bring up a spot-mic fader for a solo, drop it only 6 dB when the solo is over, not all the way off.

Often the timbre of the instrument(s) picked up by the spot mic is excessively bright. You can fix it with a high-frequency rolloff, perhaps by miking off-axis. Adding artificial reverb to the spot mic can help, too.

To further integrate the sound of the spots with the main pair, you might want to delay each spot's signal to coincide with those of the main pair. That way, the main and spot signals are heard at the same time. For each spot mic, the formula for the required delay is

$$T = D/C$$

where

T = delay time in seconds.

D = distance between each spot mic and the main pair in feet.

C = speed of sound, 1130 feet per second.

For example, if a spot mic is 20 feet in front of the main pair, the required delay is 20/1130 or 17.7 milliseconds. Some engineers add even more delay (10–15 milliseconds) to the spot mics to make them less noticeable (Streicher and Dooley, 1985).

Setting Levels

Once the microphones are positioned properly, you're ready to set recording levels. Ask the orchestra to play the loudest part of the composition, and set the recording levels for the desired meter reading. A typical recording level is +3 VU maximum on a VU meter or –3 dB maximum on a peak-reading meter for a digital recorder. The digital unit can go up to 0 dB maximum without distortion, but aiming for –3 dB allows for surprises.

When recording a live concert, you have to set the record-level knobs to a nearly correct position ahead of time. Do this during a preconcert trial recording or just go by experience: Set the knobs where you did at previous sessions (assuming you use the same microphones at this session).

Multitrack Recording

British Decca has developed an effective recording method using an eight-track recorder (Eargle, 1981, pp. 119–121):

- Record the main pair on two tracks.
- Record the distant pair on two tracks.
- Record panned accent mics on two tracks.
- Mix down the three pairs of tracks to two stereo tracks.

Stereo Miking for Pop Music

Most current pop music recordings are made using multiple close-up microphones (one or more on each instrument). These multiple mono sources are panned into position and balanced with faders. Such an approach is convenient but often sounds artificial. The size of each instrument is reduced to a point, and each instrument might sound isolated in its own acoustic space.

The realism can be greatly enhanced by stereo miking parts of the ensemble and overdubbing several of these stereo pickups. Such a technique can provide the feeling of a musical ensemble playing together in a common ambient space (Billingsley, 1989a, 1989b). Realism is improved for several reasons:

- The more-distant miking provides more-natural timbre reproduction.
- The size of each instrument is reproduced.
- Time cues for localization are included (with near-coincident and spaced techniques).
- The sound of natural room acoustics is included.

True stereo recording works especially well for these sound sources:

- Acoustic jazz combos and small folk groups (sometimes).
- Soloist or singer-guitarist.
- Drum kit (overhead).
- Piano (out front or over the strings).
- Backing vocals.
- Horn and string sections.
- Vibraphone and xylophone.
- Other percussion instruments and ensembles.

If you record several performers with a stereo pair, this method has some disadvantages. You must adjust their balance by moving the performers toward and away from the mics during the session. This is more time consuming and expensive than moving faders for individual tracks after the session. In addition, the performances are not acoustically isolated. So, if someone makes a mistake, you must rerecord the whole ensemble rather than just the flawed performance.

The general procedures for true stereo recordings follow:

1. Adjust the acoustics around the instruments. Add padding or reflective surfaces if necessary. You might prefer the sound obtained by putting the musicians near the center of a large, live room. This setup reduces early reflections but includes ambient reverberation.
2. Place the musicians around the stereo mic pair where you want them to appear in the final mix. For example, you might overdub strings spread between center and far right and horns spread between center and far left. Try to keep the acoustic bass and lead instruments or singers in the center.

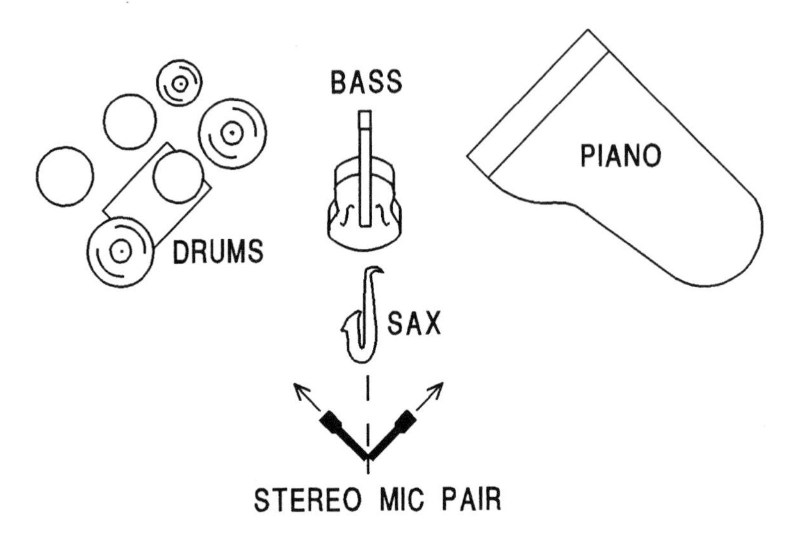

Figure 14–4 Stereo miking a jazz group.

3. Experiment with different microphone heights (to vary the tonal balance) and miking distance (to vary the amount of ambience); 3–6 feet distance is typical.
4. If some instruments or vocalists are too quiet, move them closer to the stereo mic pair, and vice versa, until the balance is satisfactory.
5. If an instrument lacks definition, consider giving it a spot mic. Mix it in at a low level.

Some excellent application notes for pop-music stereo miking are found in Billingsley (1989a, 1989b). Figure 14–4 shows a jazz group miked in stereo.

Troubleshooting Stereo Sound

Suppose that you're monitoring a recording in progress or listening to a recording you already made. Something doesn't sound right. How can you pinpoint what is wrong, and how can you fix it?

This section lists several procedures to solve audio-related problems. Read down the list of bad-sound descriptions until you find one matching what you hear, then try the solutions until your problem disappears.

Before you start, check for faulty cables and connectors. Also check all control positions; rotate knobs and flip switches to clean the contacts.

Distortion in the Microphone Signal

- Use pads or input attenuators in your mixer.
- Switch the pad in the condenser microphone, if any.
- Use a microphone with a higher "maximum SPL" specification.

Too Dead (Insufficient Ambience, Hall Reverberation, or Room Acoustics)

- Place microphones farther from performers.
- Use omnidirectional microphones.
- Record in a concert hall with better acoustics (longer reverberation time).
- Add artificial reverberation.
- Add plywood or plastic sheeting over the audience seats.

Too Detailed, Too Close, Too Edgy

- Place microphones farther from performers.
- Place microphones lower or on the floor (as with a boundary microphone).
- Using an equalizer in your mixing console, roll off the high frequencies.
- Use duller sounding microphones.
- If using both a close-up pair and a distant ambience pair; turn up the ambience pair.
- If using spot mics, add artificial reverb or delay the signal to coincide with that of the main pair.

Too Distant (Too Much Reverberation)

- Place microphones closer to performers.
- Use directional microphones (such as cardioids).
- Record in a concert hall that is less "live" (reverberant).
- If using both a close-up pair and a distant ambience pair, turn down the ambience pair.

Narrow Stereo Spread

See Figure 14–5(c):

- Angle or space the main microphone pair farther apart.
- If doing mid-side stereo recording, turn up the side output of the stereo microphone.
- Place the main microphone pair closer to the ensemble.
- If monitoring with headphones, narrow stereo spread is normal when you use coincident techniques. Try monitoring with loudspeakers or use near-coincident or spaced techniques.

Excessive Separation, Hole in the Middle, or Excessive Motion of a Soloist

See Figure 14–5(d):

- Angle or space the main microphone pair closer together, or don't use a spaced pair.
- If doing mid-side stereo recording, turn down the side output of the stereo microphone or use a cardioid mid instead of an omni mid.
- In spaced-pair recording, add a microphone midway between the outer pair and pan its signal to the center.

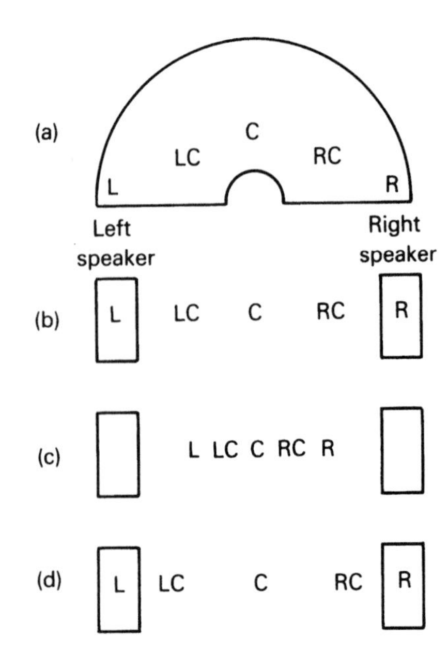

Figure 14–5
Stereo localization effects: (a) orchestra instrument locations (top view), (b) images accurately localized between speakers (the listener's perception), (c) narrow stage-width effect, (d) exaggerated separation effect.

- Place the microphones farther from the performers.
- Place the loudspeaker pair closer together. Ideally, they should be as far apart as you are sitting from them, to form a listening angle of 60°.

Poorly Focused Images

- Avoid spaced-microphone techniques.
- Use a spatial equalizer (described in Chapter 9).
- Use a microphone pair that is better matched in frequency response and phase response.
- If the sound source is out of the in-phase region of microphone pick-up, move the source or the microphone. For example, the in-phase region of a Blumlein pair of crossed figure eights is ±45° relative to center.
- Be sure that each spot mic is panned so that its image location coincides with that of the main pair.
- Use loudspeakers designed for sharp imaging. Usually, these are signal aligned, have vertically aligned drivers, have curved edges to reduce diffraction, and are sold in matched pairs.
- Place the loudspeakers several feet from the wall behind them and from side walls to delay and weaken the early reflections that can degrade stereo imaging.

Images Shifted to One Side (Left-Right Balance Is Faulty)

- Adjust the right or left recording level so that center images are centered.
- Use a microphone pair that is better matched in sensitivity.
- Aim the center of the mic array exactly at the center of the ensemble.
- Sit exactly between your stereo speakers, equidistant from them. Adjust the balance control or level controls on your monitor amplifier to center a mono signal.

Lacks Depth (Lacks a Sense of Nearness and Farness of Various Instruments)

- Use only a single pair of microphones out front. Avoid multimiking.

- If you must use spot mics, keep their level low in the mix and delay their signals to coincide with those of the main pair.

Lacks Spaciousness

- Use a spatial equalizer (described in Chapter 9).
- Space the microphones apart.
- Place the microphones farther from the ensemble.

Early Reflections Too Loud

- Place mics closer to the ensemble and add a distant microphone for reverberation (or use artificial reverberation).
- Place the musical ensemble in an area with weaker early reflections.
- If the early reflections come from the side, try aiming bidirectionals at the ensemble. Their nulls will reduce pickup of side-wall reflections.

Bad Balance (Some Instruments Too Loud or Too Soft)

- Place the microphones higher or farther from the performers.
- Move quiet instruments closer to the stereo pair and vice versa.
- Ask the conductor or performers to change the instruments' written dynamics.
- Add spot microphones close to instruments or sections needing reinforcement. Mix them in subtly with the main microphones' signals.
- Increase the angle between mics to reduce the volume of center instruments and vice versa.
- If the center images of a mid-side recording are weak, use a cardioid mid instead of an omni mid.

Muddy Bass

- Aim the bass-drum head at the microphones.
- Put the microphone stands and bass-drum stand on resilient isolation mounts or place the microphones in shock-mount stand adapters.
- Roll off the low frequencies or use a high-pass filter set around 40–80 Hz.
- Record in a concert hall with less low-frequency reverberation.

Rumble from Air Conditioning, Trucks, and So On

- Temporarily shut off air conditioning. Record in a quieter location.
- Use a high-pass filter set around 40–80 Hz. Use microphones with limited low-frequency response.

Bad Tonal Balance (Too Dull, Too Bright, Colored)

- Change the microphones.
- If a microphone must be placed near a hard reflective surface, use a boundary microphone to prevent phase cancellations between direct and reflected sounds.
- Adjust equalization. Unlike omni condenser mics, directional mics usually have a rolled-off low-frequency response and may need some bass boost.
- If strings sound strident, move mics farther away or lower.
- If the tone quality is colored in mono monitoring, use coincident-pair techniques.

References

B. Bartlett. "Microphone-Technique Basics," pp. 116–124, and "On-Location Recording of Classical Music," pp. 337–346. In *Practical Recording Techniques, Second Edition,* Boston: Focal Press, 1998.

M. Billingsley. "An Improved Stereo Microphone Array for Pop Music Recording." Preprint No. 2791 (A-2), paper presented at the Audio Engineering Society 86th convention, March 7–10, 1989, Hamburg.

M. Billingsley. "A Stereo Microphone for Contemporary Recording." *Recording Engineer/Producer* (November 1989).

J. Eargle. *The Microphone Handbook.* Plainview, NY: Elar Publishing, 1981.

S. Pizzi. "Stereo Microphone Techniques for Broadcast." Preprint No. 2146 (D-3), paper presented at the Audio Engineering Society 76th convention, October 8–11, 1984, New York.

R. Streicher and W. Dooley. "Basic Stereo Microphone Perspectives—A Review." *Journal of the Audio Engineering Society* 33, nos. 7–8 (July–August 1985), pp. 548–556.

15

BROADCAST, FILM AND VIDEO, SOUND EFFECTS, AND SAMPLING

Stereo microphone techniques have many uses besides music recording. With the advent of stereo TV (MTS) and films using Dolby Stereo or THX, the need for stereo microphone techniques has never been greater. In this chapter, we explore the application of stereo miking to broadcast, film and video, sound effects, and sampling. First, some tips that apply to all these areas.

For on-location work, if wind noise and mechanical vibration are problems, use a windscreen and a shock mount. An alternative is to use a stereo pair of omnidirectional microphones. These are inherently less sensitive to wind and vibration than directional mics. If you need an omni pair with sharp imaging, try one of the PZM boundary arrays shown in Chapter 11 or the OSS array shown in Chapter 10.

If you want sharp imaging and precise localization, try a coincident or near-coincident technique. For a diffuse, spacious sound (for example, background ambience), try a spaced-pair technique. A coincident pair or Crown SASS microphone is recommended when mono compatibility is important, which is almost always.

When the signals from an MS stereo microphone are mixed to mono, the resulting signal is only from the front-facing mid capsule. If this capsule's pattern is cardioid, sound sources to the far left and right will be attenuated. Therefore, the balance might be different in stereo and mono. If this is a problem, use an XY coincident pair rather than MS.

When compressing or limiting a stereo program, run each channel's signal through a separate channel of compression. Connect or gang these two channels of compression so that they track together. Otherwise, the image positions will shift when compression occurs.

Let's consider some stereo methods for specific uses.

Stereo Television

In stereo TV applications, stereo mic techniques are used mainly for electronic news gathering, audience reaction, sports, and classical music.

Imaging Considerations

Stereo TV presents some problems because of the disparity between picture and sound. Some viewers will be listening with speakers close to either side of the TV; others will be listening over a stereo system with speakers widely spaced. Where should the images be placed to satisfy both listeners? Should the sound-image locations coincide with the TV's visual images?

Extensive listening tests (Rumsey, 1989, pp. 13, 14, 41) yielded these findings:

- Listeners prefer dialogue to be mono or to have only a narrow stereo spread. A two-person interview with dialogue switching from speaker to speaker is distracting.
- Listeners can accept off-camera sounds that originate away from the TV screen because the TV screen is considered a window on a larger scene.
- Listeners prefer sound images to be stable, even if the picture changes.
- Listeners easily notice a left-right reversal because the sound images don't match those on the screen.

Based on these results, many broadcasters recommend these practices for stereo TV:

1. Record dialogue in mono in the center or with a narrow stereo spread.
2. Record effects, audience reaction, and music in stereo. You can allow off-camera sounds to be imaged away from the TV screen.

3. To prevent shifting images, don't move (pan) a stereo microphone once it is set up.
4. Avoid extreme differences between sound and picture. Be careful not to reverse left and right channels (say, by inverting an end-fired stereo microphone).

Mono-Compatibility

An important requirement for any stereo broadcast is mono-compatibility. Since many TV viewers are still listening in mono, the audio signal must sound tonally the same in mono or stereo. Also, excessive L – R signals in noncompatible programs can cause distortion and image shifting in AM stereo (Pizzi, 1984).

Phasing problems can arise when stereo signals with interchannel time differences are mixed to mono. Various frequencies are canceled, causing a hollow or dull tone quality. Phase cancellations in mono also disrupt the relative loudness (balance) of instruments in a musical ensemble and change the volume of different notes played on the same instrument.

To ensure mono-compatibility, use a mono-compatible microphone array, such as a coincident pair, MS stereo microphone, or Crown SASS microphone. Also, when you use multiple close microphones panned for stereo, be sure to follow the 3:1 rule (Pizzi, 1984). The distance between microphones should be at least three times the mic-to-source distance (as shown in Figure 15-1). This prevents phase cancellations if the stereo channels are summed to mono.

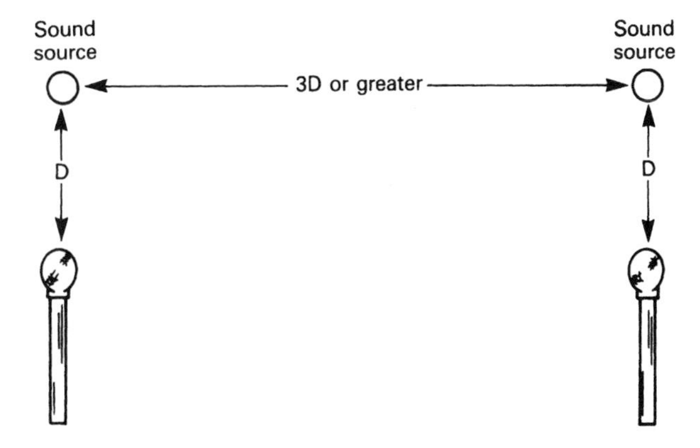

Figure 15-1 The 3:1 rule of microphone placement avoids phase interference between microphone signals.

Monitoring

Although headphones often are used for monitoring on location, they can give a very different impression of stereo effects than loudspeakers. For this reason, it's wise to monitor in a control room with small close-field monitor loudspeakers. These are placed about 3 feet from the listener to minimize the influence of room acoustics. Include a stereo-mono switch in the monitoring chain to listen for the problems described earlier (Pizzi, 1984).

How far apart should monitor speakers be for stereo TV listening? A typical TV viewer sees the TV screen covering a 16° angular width, while the optimum stereo listening angle for speakers is 60°. The disparity between these two angles can be distracting. A speaker angle of 30° (as shown in Figure 15-2) seems to be an adequate compromise for stereo TV listening ("The Link," 1984; Lehrman, 1984). Keep the speakers at least 1 foot from the video monitor to prevent magnetic interference.

Electronic News Gathering

The current practice in electronic news gathering (E.N.G.) is to pick up the correspondent's commentary in mono with a close-up handheld or lavalier

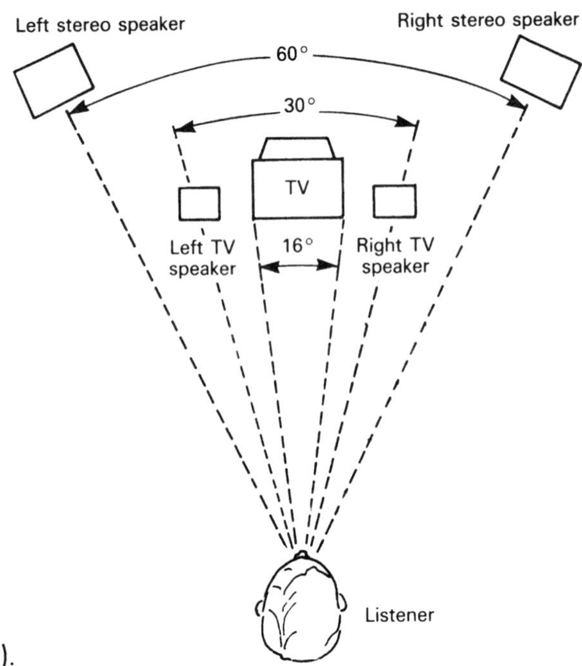

Figure 15-2
TV viewing angle (16°), stereo music listening angle (60°), and compromise stereo-TV listening angle (30°).

mic. This signal is panned to center between the loudspeakers. Stereo commentary is distracting: It is visually confusing to hear commentary to the left or right of the TV picture, removed from the image of the person speaking. Background ambience, however, is recorded in stereo for added realism. The mono commentary usually is mixed live with the stereo ambience.

Commentary and stereo ambience also can be mixed in postproduction by taping the announcer and stereo background on a multitrack recorder (Pizzi, 1983). If you use an M2 format VTR, however, this method can cause phasing problems. The M2 format recorder contains four audio tracks: two FM and two linear with Dolby C. The FM tracks are locked to speed, but the linear tracks are always shifting in time. Therefore, the two stereo pairs of tracks continuously shift in phase relative to each other. For this reason, you may want to mix live if you use the M2 format.

A useful tool for on-location E.N.G. is a handheld stereo microphone. Currently, models of this type are made by Audio-Technica, AKG, Crown Sony, and Sanken. The Sanken must be used with a shock-mount pistol grip. These microphones can be used just to pick up background sounds or both commentary and background with a single microphone. Figure 15-3 shows both types of E.N.G. pickup.

In the latter case, the reporter speaks directly in front of the microphone in the center. This results in a mono signal. The closer you place the mic to the reporter, the less background noise you pick up. But if background noise is excessive—for example, at a fire or a rock concert—the dialogue may not be intelligible with this method.

Audience Reaction

It is common in talk shows, game shows, or plays to pick up dialogue in mono with close mics and cover audience reaction and music in stereo. Hearing the audience in stereo greatly enhances the feeling of being there. The close mics can either be standard mono units or MS stereo units.

Stereo audience reaction can be picked up in many ways:

- With boundary mics on the stage front, walls, ceiling, or balcony edge.
- With multiple mono microphones hung over the audience, panned for stereo.
- With a single stereo microphone or mic pair.
- With a combination of stereo mics and panned mono mics (Figure 15-4).

If the audience mics are far from the stage (say, 30 feet or more), their signal will be delayed relative to the on-stage mics. So you need to delay

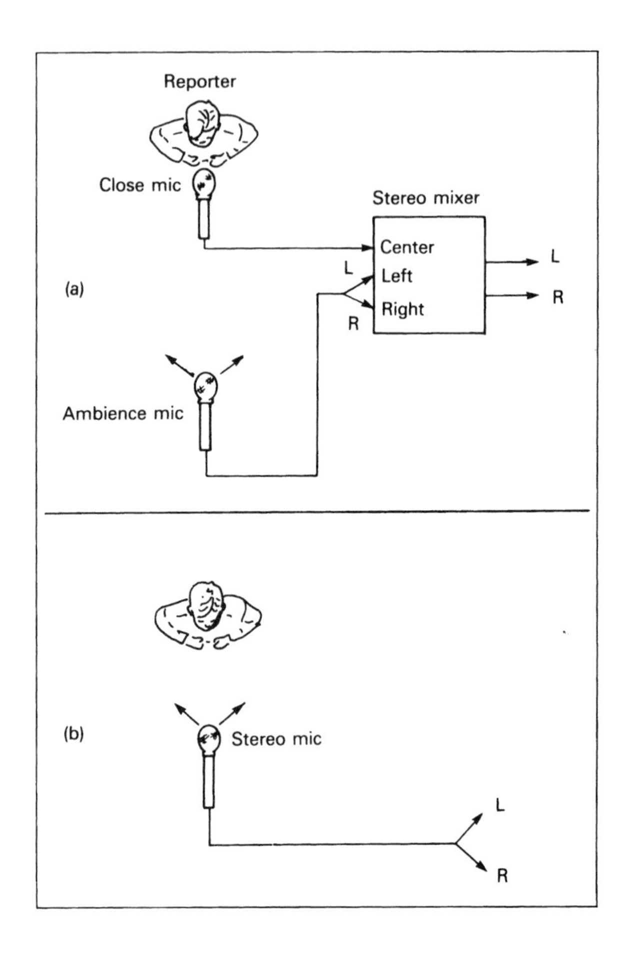

Figure 15-3
Typical stereo miking arrangements for E.N.G.: (a) a spot mic for the reporter plus a stereo mic for ambience, (b) a stereo mic picking up the reporter and ambience.

the stage mics (with a delay unit or digital editing software) so that their signal coincides with the audience mics.

If you're miking the audience with a pair of mics close to the stage (to reduce the delay just mentioned), place the mics up high and aim them at the back row of the audience.

Spaced microphones give a desirable spacious feeling to audience reaction because of random phase relationships between channels.

If the audience reaction is clearly imaged, it might be so realistic as to be distracting. For this reason, some engineers prefer to mix the audience mics to mono and run the mono mix through a stereo synthesizer.

If actors play near only one side of the audience, most of the reaction will be on that side. This one-sided reaction, left or right, may be confusing to home listeners. To hear this reaction on both sides in stereo, you may need to set up a stereo mic pair for each section of the audience.

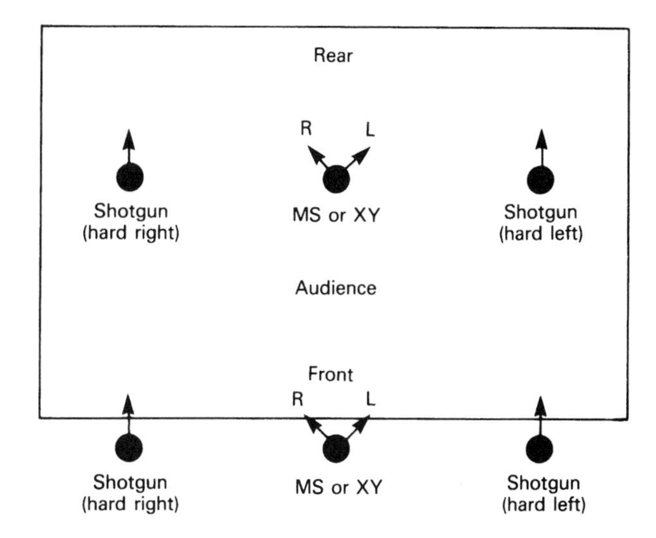

Figure 15-4 Stereo audience miking used for the *David Letterman Show.*

The following tips on audience miking are from an article by Shawn Murphy (1986). The larger the audience, the fewer audience-reaction mics you need. Two or three mics can cover an audience of 3000, while eight mics might be needed for an audience of 200.

It is important that the sound of the audience reaction match the sound of the dialogue. This is more likely when you use the same type of microphone for both sound sources. When you turn dialogue mics and audience mics on and off, the only change you should hear is an increase in the level of the PA system. Try to position the microphones to not pick up the sound-reinforcement speakers.

A typical signal-processing chain for the audience mics is as follows, in this order:

1. A stereo submixer to mix the mics to two channels.
2. A foot pedal for volume control.
3. A 100 Hz high-pass filter to reduce room rumble.

Parades

Picking up parades clearly in stereo is difficult because of background noise. However, certain special stereo microphones provide a tighter sound with less background pickup than standard stereo microphones. For example, Neumann offers an MS stereo shotgun microphone (see

Chapter 16). The side element is a bidirectional capsule; the middle element is a short shotgun microphone. This microphone has been used to pick up parades in stereo. The Pillon stereo PZM (Chapter 11) is another "tight sounding" microphone pair.

To pick up marching bands at the Mummer's Parade on New Year's Day in Philadelphia, KYW-TV3 used a pair of supercardioid boundary microphones placed back-to-back on a 2-foot-square piece of plexiglass. This arrangement was found to work better than conventional mics and shotguns. To reduce wind noise, the plexiglass boundary was wrapped in acoustically transparent fabric. This array was suspended on cables 32 feet over the street and 24 feet in front of the performance line of the bands (as shown in Figure 15-5).

To pick up the band as it approached the judges' stand, a similar array was mounted on a boom standing in the middle of the groups' performance area. Two backup mics also were mounted on the same boundaries (Crown International, 1988).

Shotgun microphones often are needed in outdoor applications for isolation from the PA and background noise. Shotgun mics are preferred over parabolic microphones for sideline pickup of parades, because parabolics have a narrowband frequency response unsuitable for music.

Sound engineer Ron Streicher has used two shotguns crossed in an XY pattern for stereo pickup. He high-pass filters the shotguns to reduce wind noise and mixes them with a single low-pass-filtered omni microphone to pick up low frequencies.

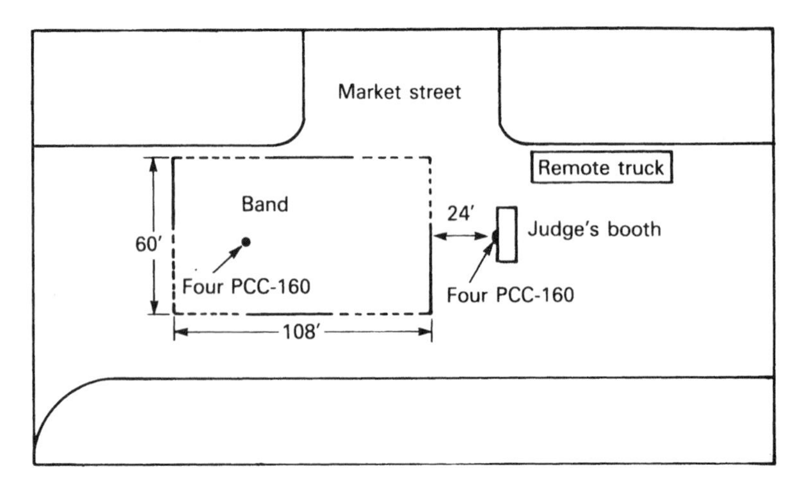

Figure 15-5 Stereo pickup of a marching band in a parade.

Sports

Sporting events include three basic sound sources:

- The crowd reaction (including ambient acoustics).
- The sounds of the sport itself: players' yelling, baseball bat hits, tennis ball hits, bowling pin crashes, wrestling grunts, and so on.
- The announcer's commentary.

For crowd reaction and ambience, use an overall coincident pair or stereo microphone. If you use more than one such microphone, separate them widely. An alternative is to use widely spaced omnis or boundary mics. Widely spaced mics give poor image focus and a ping-pong effect but are more mono-compatible than closely spaced microphones.

For sport sounds, use close-up spot mics panned as desired. These mics usually are shotguns or parabolic reflectors. Mini cams can be equipped with stereo microphones, which are switched to follow the video shot.

For the announcers, use headset mics. A single announcer can be panned to center; two can be panned lightly left and right. Another method is to pan play-by-play commentary to center; pan color #1 partly left and color #2 partly right. Roving reporters can use handheld wireless mics panned to center or panned opposite a booth announcer (Pizzi, 1983).

Sport sounds often require custom microphone pairs. Basketball games can be picked up with a stereo pair over center court, such as two boundary mics back to back on a panel, or boundary mics on the floor at the edge of the court and on the backboards under the hoop. Indoor sports such as weight lifting or fencing can be picked up with a near-coincident pair of directional boundary microphones or an MS boundary pair on the floor. For bowling, try a boundary mic on the back wall of the alley, high enough to avoid being hit, to pick up the pin action. Use a stereo pair on the alley itself (Crown International, 1999).

A baseball game can be covered as suggested by Pizzi (1983; see Figure 15-6). Use a stereo mic near home plate, mixed with widely spaced boundary mics for ambience, plus shotgun or parabolic mics near the foul line aiming at the bases. Outfield microphones (cardioids or shotguns) are optional.

Multiple-sport events such as the Olympics can be covered with a number of MS mics or SASS mics at different locations. These mics pick up scattered events throughout the stadium or field.

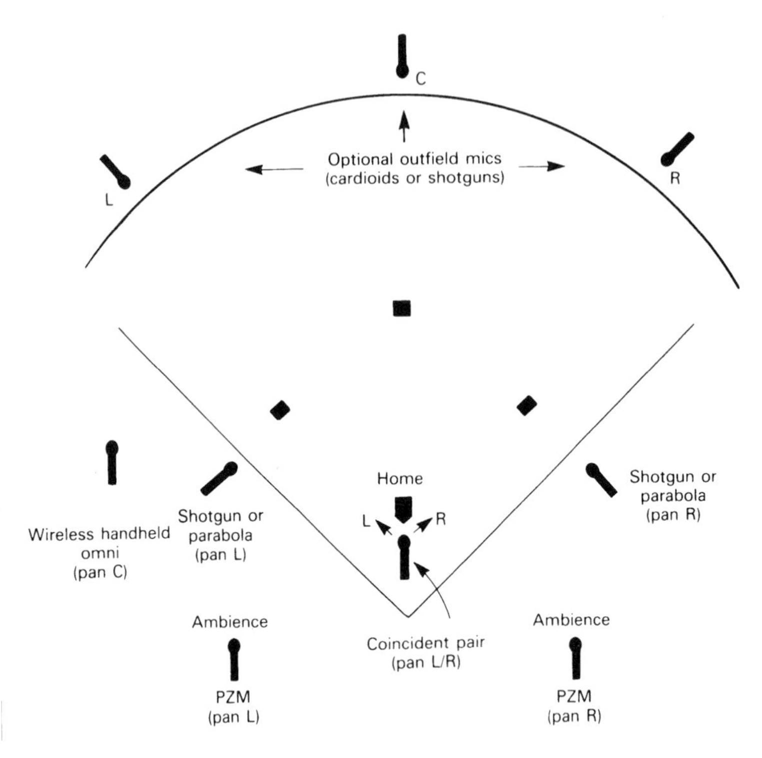

Figure 15-6 A suggested stereo microphone setup for baseball (after Pizzi).

Stereo Radio

Stereo mic techniques are also needed for radio group discussions and plays.

Radio Group Discussions

For a group of people seated halfway around a table, try an XY pair (or a stereo microphone set to XY) over the center of the table (Figure 15-7). This will localize the images of the participants at various locations. To prevent phase cancellations due to sound reflections off the table, cover it with a thick pad or use an MS boundary microphone (described in Chapter 5)—or just remove the table.

When you broadcast a one-on-one interview, try to have the voices in mono and the ambience in stereo, because ping-pong interviews can be distracting. To do this, use an MS microphone and set the middle microphone to a bidirectional pattern. Place the participants on opposite sides of

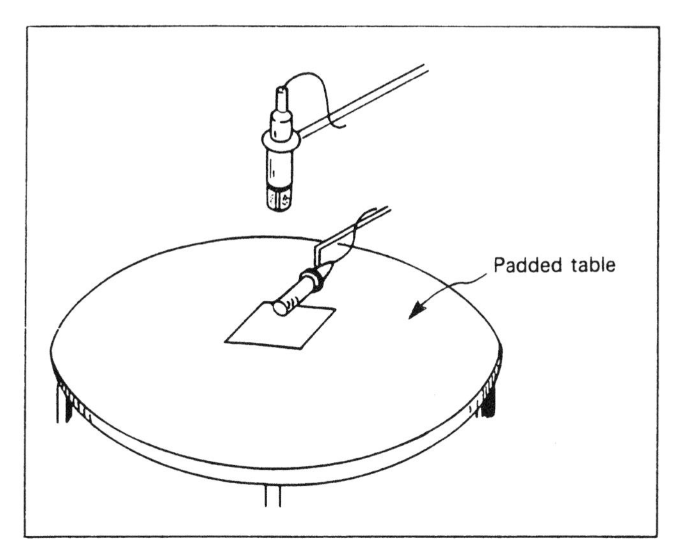

Figure 15-7 Picking up a radio group discussion with an XY stereo microphone over the center of the table or with an MS boundary microphone on the table.

the middle microphone (front and rear). Control the amount of ambience by adjusting the side microphone level (Pizzi, 1983). An alternative is to use a Blumlein array (crossed figure eights) with the participants seated at the front and rear of the array.

Radio Plays

Try a coincident pair, stereo microphone, or dummy head (for headphone reproduction) aiming at the center of the action. The actors can simulate distance by moving away from the microphones. Good-sounding positions can be marked with tape on the floor.

It is important that the actors don't go beyond the pickup angle of the stereo mic pair. This is especially true for the Blumlein pair of crossed figure eights, because the left and right channels will be in opposite polarity if the actor is past the axis of either microphone.

To determine the stereo pair's pickup angle, monitor the output of the pair in stereo while someone speaks in front of it. Have that person start in the center and slowly move to one side while speaking. When the monitored image of the person speaking is all the way to one speaker, have the person stop and mark the floor with tape at that point. Do the same in the other direction off center. Tell the actors to stay within the two tape marks.

You may want to supply the actors with stereo headphones so that they can hear the effects of their position and closeness to the mics. However, note that headphone monitoring of a coincident pair has little stereo separation.

Film and Video

Stereo miking also finds use in feature films, as well as video documentaries and industrial films.

Feature Films

Sound engineers for feature films usually pick up dialogue with one or two shotgun mics or lavalier mics to reduce ambience. During postproduction, the mono signals from these microphones are panned to the desired stereo locations to match the visual locations of the actors on or off screen. This forms a dialogue premix. Often the original dialogue track recorded on location is unusable, so it is replaced in the looping process.

The final soundtrack is a stereo mix of several stereo premixes:

1. Dialogue premix.
2. Sound effects premix.
3. Ambience or atmosphere premix.
4. Music score premix.

An alternative to panning is to record dialogue in stereo with a stereo shotgun mic on a boom, a PZM stereo shotgun, or an SASS microphone. The stereo perspective of the sound should match the visual perspective on a theater screen.

You might want to add a stereo mic or stereo pair for ambience. The realism of background ambience or wild sound is enhanced by stereo recording.

Sennheiser has suggested a novel MS stereo miking method for films, TV, and so on: The side element of the MS array is a stationary side-aiming bidirectional microphone; the middle element is a boom-mounted shotgun mic that you aim at the sound source. The shotgun is physically panned to follow the action (Gerlach, 1989).

Documentaries and Industrial Productions

For these productions, you can either pick up both dialog and ambience with a single stereo microphone (handheld or stand mounted) or use a close-up mic for the announcer, mixed with a stereo array for ambience (as shown in Figure 15-3).

In positioning the stereo microphone, hold it up high at arm's length above the camera, aiming down at the sound source, or hold the mic below the camera, aiming up at the sound source. A stereo microphone can be attached to a Steadicam platform to follow the action. The first use of this method was by Gary Pillon of General Television Network, Oak Park, Michigan. Using a stereo PZM microphone that he designed, he recorded the soundtrack for a documentary that subsequently won an Emmy. Now he uses an SASS mic.

Another way to pick up stereo ambience is to use a coincident pair of short shotgun microphones. They produce a wide stereo effect with a hole in the middle. This hole can be filled with the announcer's voice in mono (Pizzi, 1983).

For outdoor ambience, you might try a pair of omni microphones spaced more than 25 feet apart. Since the coherence between microphones is small at that spacing, mono cancellation is reduced. Also, the omni mics are relatively insensitive to wind noise (Pizzi, 1983).

Sound Effects

Recording sound effects in stereo requires some special considerations. When you record moving sound sources, such as truck or plane passes, you need a stereo mic array that accurately tracks the motion. Coincident and near-coincident techniques work well in this application. With spaced-microphone recordings, the sound image usually jumps from one speaker to the other as the sound source passes the mic array. Accurate motion tracking in sound-effect recording is analogous to accurate localization in music recording.

When recording a moving effect, experiment with the distance between the microphone and the closest pass of the sound source. The closer the microphone is to the path of the subject, the more rapidly the image will pass the center point (almost hopping from one channel to the other). To achieve a smooth side-to-side movement, you may need to increase the distance (Crown International, 1999).

Since you'll be monitoring your recording in the field with headphones, you may want to use a stereo miking technique that provides the same stereo spread on headphones as on loudspeakers. Recommended for this purpose are near-coincident arrays, such as ORTF, NOS, OSS, or the SASS microphone.

For recording animals, some recordists hide the microphones in shrubbery near the animals and run long cables back to the recorder so they won't inhibit the animals.

An award-winning nature recordist, Dan Gibson, has recorded soundtracks for wildlife films and for the Solitudes series of records. He records digitally with a Sony DAT recorder. For birds and long-range pick-up, Gibson uses a stereo parabolic microphone P200S, for mid-range distances he uses a pair of Neumann shotgun mics, and for close-up work he uses a pair of AKG D 222 microphones.

To edit out noises in his DAT recordings, he first makes a copy from one DAT to another. He plays the first recording up to the noise, then quickly cross-fades to the other recording cued up just after the noise.

Some miscellaneous tips on sound-effect recording:

- Use a shock mount and windscreen outdoors.
- If you want to reduce pickup of local ambience and background noises, place the mic close to the sound source.
- Record flat; most effects need EQ later.
- Record multiple takes so that you can choose the best one.

Sampling

Sampling is the process of digitally recording a short sound event, such as a single note from a musical instrument. The recording is stored in computer memory chips. You play it back by pressing keys on a piano-style keyboard. The higher the key, the higher is the pitch of the reproduced sample.

Stereo sampling allows recording on two channels for stereo localization. If you play, for instance, a stereo piano sample, each note is reproduced in its original spatial location.

When sampling, be sure to use microphones with low self-noise (for example, less than 20 dBA). If the sample is noisy, the noise will audibly shift in pitch as you play different keys on your sampling keyboard. Variable-pitch noise is easier to hear than steady noise.

Another requirement for a stereo sampling microphone is a wide, flat frequency response. This avoids tonal coloration of the sample. A hypercardioid or supercardioid pattern, as well as close microphone placement, can help to reduce pickup of ambient noise and room reflections, so that you hear only the desired sound source.

When you record samples or sound effects for keyboard, drum-machine, or disk-soundbank reproduction, any recorded ambience will be reproduced as part of the sample. For added future flexibility, you may want to make several samples of one source at different distances to include the range of added reverberance.

Off-center images can be reproduced accurately by sampling the sound source in the desired angular position as perceived from stereo center. Recorded ambience will sharpen the image, but is not necessary.

When you pitch-shift a stereo sample, the image location and the size of the room will change if you use spaced or near-coincident techniques. That is because the interchannel delay varies with the pitch shift. To minimize these undesirable effects, try sampling at intervals of one-third octave or less or record the sample with a coincident-pair technique (Crown International, 1999).

When looping, try to control the room ambience so that it is consistent before and after the sample (unless reverberant decay is desired as part of the sample).

References

Crown International. *Mic Memo* (April 1988). Elkhart, IN: Crown International, 1988.

Crown International. Microphone catalog. Elkhart, IN: Crown International, 1999.

Crown International. *Crown Boundary Microphone Application Guide.* Elkhart, IN: Crown International, 1999. (The material for the sampling portion was provided by Michael Billingsley.)

H. Gerlach. "Stereo Sound Recording with Shotgun Microphones." *Journal of the Audio Engineering Society* 37, no. 10 (October 1989), pp. 832-838.

P. Lehrman. "Multichannel Audio Production for US TV." *Studio Sound* (September 1984), pp. 34–36.

"The Link." *High Fidelity* (November 1984).

S. Murphy. "Live Stereo Audio Production Techniques for Broadcast Television," pp.175–179. *The Proceedings of the American Engineering Society Fourth International Conference on Stereo Audio Technology for Television and Video,* May 5–18, 1986, Rosemont, IL.

S. Pizzi. "A 'Split Track' Recording Technique for Improved E.N.G. Audio." Preprint No. 2016, paper presented at the Audio Engineering Society 74th convention, October 1983.

S. Pizzi. "Stereo Microphone Techniques for Broadcast." Preprint No. 2146 (D-3), paper presented at the Audio Engineering Society 76th convention, October 8–11, 1984, New York.

F. Rumsey. *Stereo Sound for Television.* Boston: Focal Press, 1989.

My special thanks to Terry Skelton, staff audio instructor at NBC, for his helpful suggestions on this chapter.

16

STEREO MICROPHONES AND ACCESSORIES

This chapter lists stereo microphones, dummy heads, MS matrix boxes, and stereo microphone stand adapters. The list is up to date only for the date of publication of this book. Since models and prices will change, please contact the manufacturers for current information. This chapter should be viewed less as a catalog and more as an illustration of what features are available. Prices range from $500 to $6735.

The following terms are used in microphone specs. The axis of maximum sensitivity in *side-addressed* mics is at right angles to the microphone's long axis. You aim the side of the mic at the sound source. The axis of maximum sensitivity in *end-addressed* mics is the same as the mic's long axis. You aim the front of the mic at the sound source.

Stereo Microphones

- **AKG C-426B:** Two twin-diaphragm condenser capsules, one atop the other, for MS or XY use. Three patterns plus six intermediate steps. Low-cut switch, attenuator, and shock mount. Side addressed. See Figure 16–1.
- **Audio-Technica AT822:** End-addressed XY mic. Compact and lightweight for DAT recording, TV, FM field applications, and camera mount. Switchable low-frequency rolloff, battery powered only, unbalanced outputs. Several accessories.

Figure 16–1 AKG C-426B stereo microphone. (Courtesy AKG Acoustics, Inc.)

- **Audio-Technica AT825:** Same as preceding but with battery or phantom power, balanced outputs, and flatter response.
- **Audio-Technica ATR25:** End-addressed XY mic designed for video cameras. On-off switch. Several mounts and cables.
- **Beyerdynamic MC 742:** Two double-diaphragm condenser capsules mounted one atop the other. Capsules can be rotated 360°, and the five selectable polar patterns suit the mic for MS or XY recording. Side addressed. An MSG 740 remote control-power unit is available.
- **Beyerdynamic MC 833:** Three large-diaphragm capsules with variable positions. Works for MS or XY recording without an external

matrix. For orchestra or recital recording, ambience, and sampling. End addressed.

- **Beyerdynamic MCE 82:** Two cardioid capsules at 90° for XY stereo. Low cut, battery/phantom, many accessories. For remote recording, E.N.G., sound reinforcement, studio recording. End addressed.
- **Crown SASS-P MKII PZM Stereo Microphone:** Two ear-spaced PZMs on angled boundaries. For near-coincident, mono-compatible recording, E.N.G., and the like. End addressed.
- **Neumann SM69 Fet:** Multipattern remote-controlled dual stereo MS/XY microphone. Upper capsule can be rotated through 270° relative to the lower one. Side addressed. See Figure 16–2.
- **Neumann USM 69i:** Same as Neumann USM SM69 fet, but with directional pattern switches located on the mic itself, eliminating the need for the in-line box. Side addressed.
- **Neumann RSM 191:** Stereo shotgun XY mic. Mono-compatible. The middle mic is a hypercardioid short shotgun, while the side capsule system has a figure-eight pattern. An amplifer-matrix unit offers XY stereo with varying acceptance angles or MS outputs with multliple ratios. External phantom power or 9V battery inside the matrix box. End addressed. See Figure 16–3.
- **Neumann KFM 100:** A 20 cm (7.8 inch) wooden sphere with two diametrically opposed pressure mics flush mounted in the sphere. Time and spectral differences between channels produce the stereo imaging, with excellent depth reproduction. Response down to 10 Hz. Side addressed.
- **Pearl DS60:** Two rectangular dual-membrane capsules mounted one above the other, 90° apart. For XY or MS methods. The required polar pattern is chosen at the console: cardioid, figure-eight, omni, XY, MS, or Blumlein. Phantom powered. Four XLR connectors. Side addressed.

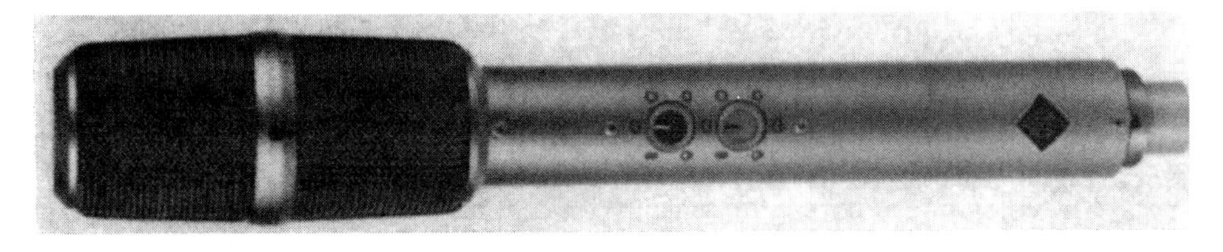

Figure 16–2 Neumann SM69 fet stereo microphone. (Courtesy Neumann USA.)

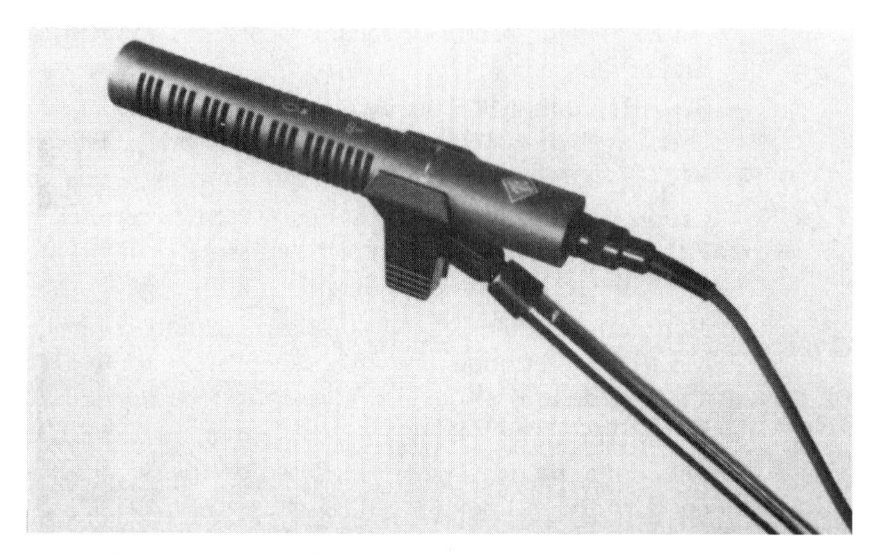

Figure 16–3 Neumann RSM 191 stereo shotgun microphone. (Courtesy Neumann USA.)

- **Pearl TL4/TL44:** For mono or stereo use. Two discrete cardioid outputs may be used together or apart to get cardioid, omni, figure-eight, or 180° XY via the mixing console. Phantom powered. Side addressed.
- **Pearl MS2/MS8:** Lightweight mics for MS recording in TV, video, and film production. In the MS2, the M and S signals are combined by an internal matrix to produce left and right outputs. In the MS8, the M and S channels are projected independently for external processing, such as recording followed by an MS matrix. Phantom powered. Side addressed.
- **Sanken CMS-2:** Small, lightweight, handheld MS mic for TV, film, and music. Cardioid middle element and bidirectional side element. End addressed. See Figure 16–4.
- **Sanken CMS-7:** Portable MS mic for broadcast and film. Requires external matrix box. Fixed angle of polar patterns. CMS-7 has cardioid middle unit; CMS-7H has hypercardioid middle unit. End addressed. See Figure 16–5.
- **Sanken CMS-9:** Portable MS mic similar to CMS-7 but with an internal matrix so no external box is required. Left-right or MS outputs, fixed angle of polar patterns. End addressed.

Figure 16–4
Sanken CMS-2 stereo microphone. (Courtesy Pan Communications, Inc.)

- **Sanken CSS-5:** Shotgun stereo microphone with three patterns: mono short shotgun, normal XY stereo, and wide XY stereo. End addressed.
- **Schoeps KFM6 Sphere Microphone:** Two omni capsules flush-mounted in an 8 inch wooden globe. Head-related phase and frequency compensation. Phantom powered. Side addressed.
- **Schoeps MSTC 64 "ORTF" Microphone:** Two MK 4 cardioid capsules at either end of a T-shaped dual amplifier body, spaced 17 cm (6.7 inches) apart and angled 110°. XLR-5M output, 12–48V phantom powered.
- **Schoeps CMXY 4V:** Two side-addressed cardioid CCM 4Vg capsules, interlocked so that they rotate in opposite directions to adjust stereo spread from 0° to 180°. Compact, with swivel base for hanging, boom mounting, or tabletop use. 12–48V phantom powered.

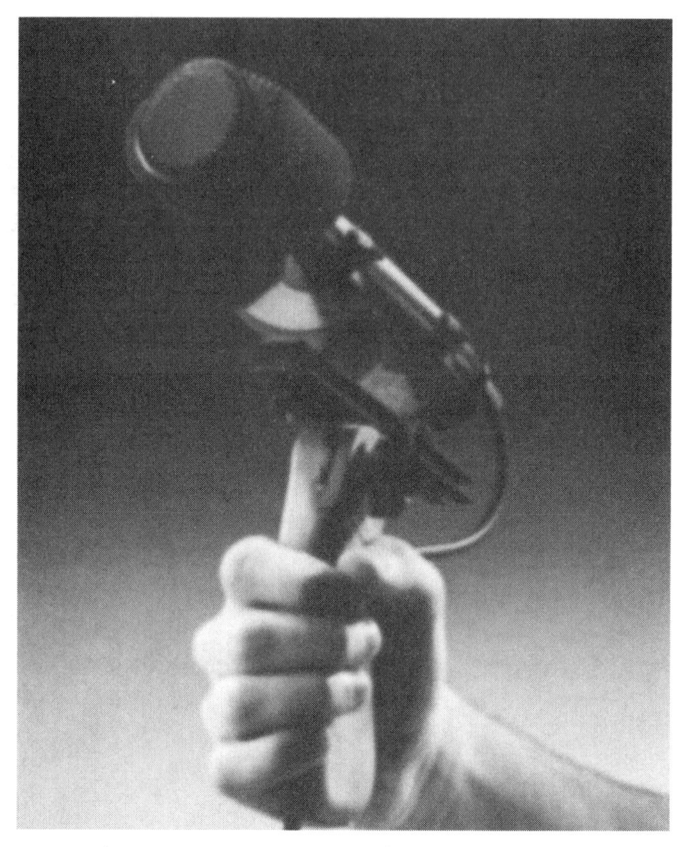

Figure 16–5 Sanken CMS-7 stereo microphone. (Courtesy Pan Communications, Inc.)

Choice of connectors. Two CMBI/MK or CMC/MK mics, with Schoeps accessories, can be set up for compact XY, MS, or near-coincident pickups.

- **Shure VP88:** MS mic with stereo or MS outputs. Switchable stereo spread control on microphone. Low-cut switch. End addressed.
- **Sony ECM999:** Compact, lightweight, end-addressed MS mic for recording or news gathering. Battery-check LED, variable stereo angle, low-cut filter. Battery powered.
- **Sony ECMMS5:** As preceding, but with six-position stereo-angle selector. Phantom power or battery operation with optional DC-MS5.
- **SoundField Microphone (ST250, MKV):** Uses four capsules arranged in a tetrahedron, phase matrixed for true coincidence. Processor offers remote control of polar pattern, azimuth (horizontal

rotation), elevation (vertical tilt), and dominance (fore and aft movement). Separate outputs for stereo and ambisonic (3-D) surround sound. Side addressed. With flight case and 20 meter cable.

- **SoundField SPS422:** Less-expensive SoundField model. Two forward-facing capsules (left front and right front) and two backward-facing capsules (left back and right back). Processor offers remote control of pattern and width.
- **SoundField 5.1 Microphone System:** A single, multiple-capsule microphone (SoundField ST250 or MKV) and SoundField Surround Decoder for recording in surround sound. The decoder translates the mic's B-format signals (X, Y, Z, and W) into L, C, R, LR, RR, and mono subwoofer outputs.

Dummy Heads

- **Aachen Head Model HMS II:** Dummy head with omni measurement microphones. A record processor equalizes the head signals to have flat response in a frontal free field. A reproduce unit contains free-field equalizers for use with Stax SR-Lambda Professional headphones. Also available for recording is a unit with Schoeps capsules, two XLR-type outputs, and switchable EQ for frontal free-field or random-incidence sound field.
- **Bruel & Kjaer 4100 Head and Torso Simulator:** Uses two B&K mic capsules in a detailed replica of a human head and torso.
- **Core Sound:** This company offers a variety of binaural microphone sets that you wear on your head. Omni or cardioid capsules. Powered by battery box or DAT recorder.
- **Holophonics System:** "Does not use traditional microphones. Instead, a proprietary sound processing technique is used that captures the full spectrum of essential information traveling from the ear to the the brain in the recording environment."
- **Knowles Electronics KEMAR Dummy Head:** Designed for research rather than dummy-head recording.
- **Neumann KU 100 "Fritz III" Dummy Head Binaural System:** A detailed human-head replica with omni microphones inside of the ears. Loudspeaker compatible. For music recording, radio drama, film special effects, outdoor nature recordings, acoustic evaluation, and scientific research. Powered by internal batteries or external phantom power supply. See Figure 16–6.

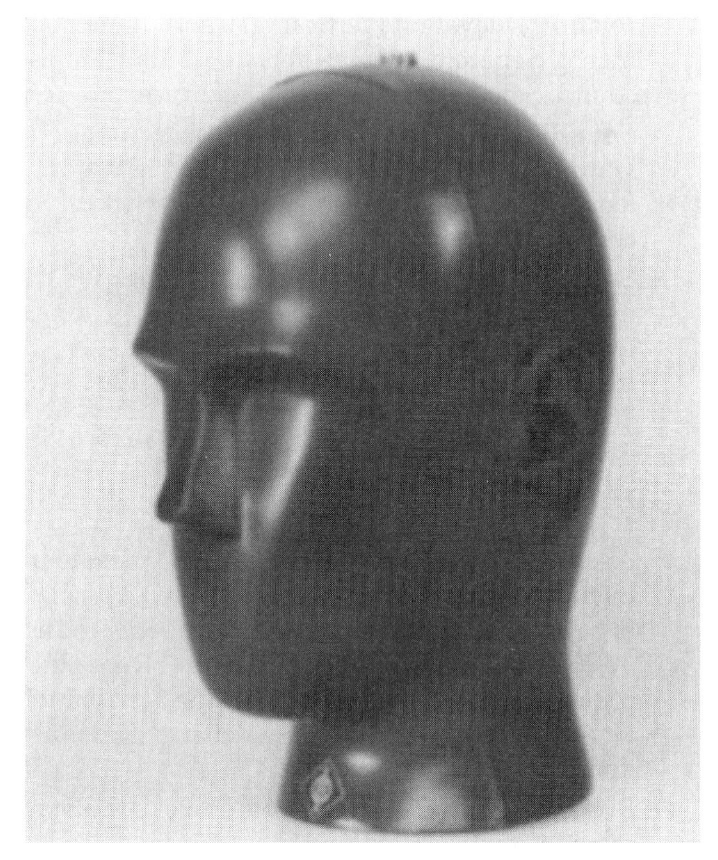

Figure 16–6 Neumann KU 100 dummy head. (Courtesy Neumann USA.)

- **Sonic Studios Dimensional Stereo Mics (DSM):** Two mini omni condenser mics meant to be worn on your head (just forward of the ears). Mics can be mounted on the DSM-GUY dummy head, which is made of Sorbothane. Powered by Sony recorders that have plug-in power or by external powering adapters. Not equalized for head diffraction. Several DSM models are available.

Stereo Microphone Adapters (Stereo Bars)

- **Audio Engineering Associates (AEA) Stereo Microphone Positioner:** Positions coincident and near-coincident arrays. Handles large mics such as Coles 4038. Vertical or horizontal arrays, stand mounted or hung. Rotation angles are lines engraved at ±30°, 45°, and 55°. Center-to-center spacing is marked on ruler with ORTF position shown.

- **AEA Tree Series:** Mini-tree, Decca tree, and large tree multiple microphone array mounts are used to securely position heavy microphones. The AEA modular microphone array can be configured to hold microphones for 5.1 channel recording.
- **AKG KM-235/1:** V-shaped adjustable stereo bar or rail.
- **AKG H10:** Metal stereo crossbar with two $\frac{3}{8}$-inch knurled-head screws. Can be spaced $1\frac{5}{8}$–3 inches apart.
- **AKG H52:** Stereo suspension set for condenser mic capsules CK1 X or XK 3 X. For mounting on mic stand or hanging from ceiling. For coincident or near-coincident arrays.
- **Beyer ZMS1:** V-shaped adjustable mounting rail. Fits two mics for stereo recordings with a maximimum separation of 8 inches.
- **Bruel & Kjaer CXY4000:** Adjustable linear stereo rail for XY, MS, near-coincident, or spaced-pair miking.
- **Neumann DS21 Dual Microphone Mount:** Can be used to combine two miniature microphones and two bent-capsule extension tubes into one fixed assembly for stereo recordings. See Figure 16-7.
- **Sanken XY Stereo Microphone Mount**: Used to set a pair of CU-41 microphones with S-41 adapters for XY stereo recording.
- **Schoeps UMSC:** Universal stereo bracket with ball joint. Has two sliding, locking clamps on a crossbar, with detents set at the correct angles and positions for XY, MS, and ORTF. Aligns two MK capsules. See Figure 16–8.

Figure 16–7
AK20 (bidirectional) and AK40 (cardioid) capsules with the DA-AK elastic suspension for MS use. (Courtesy Neumann USA.)

Figure 16–8 Schoeps UMSC Universal stereo bracket. (Courtesy Posthorn Recordings.)

- **Schoeps STC:** ORTF T-bar with SG20, sets two cardioid capsules 17 cm (6.7 inches) apart and angled 110°.
- **Schoeps MS-BLM:** Boundary-layer twin clamp for MS, mounts KC5 and MK8 on a BLM-3 plate.

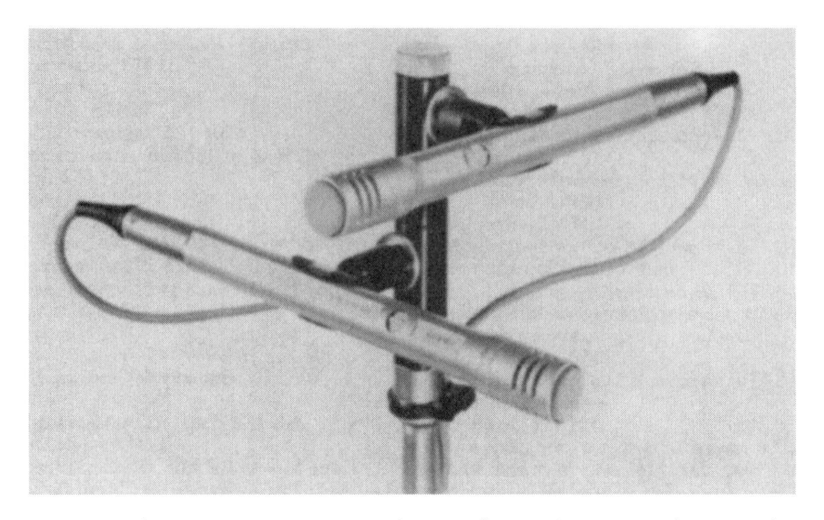

Figure 16–9 Shure 27M stereo microphone adapter. (Courtesy Shure Brothers, Inc.)

- **Schoeps SGMSC:** Mounting stud for MS setup of MK8 and MK-. With swivel, stand clamp, $\frac{3}{8}$-inch adapter.
- **Schoeps AMSC:** Elastic suspension with swivel. Holds MK8 and MK-mic capsules parallel for mid-side pickup.
- **Shure A27M:** Uses two rotating stacked cylinders. Lets you adjust angle and spacing for coincident and near-coincident methods. See Figure 16–9.

MS Matrix Decoders

- **Audio Engineering Associates MS38 Dual-Mode Active MS Matrix:** Input switchable MS or left-right. Adjustable mid-side ratio (stereo spread control).
- **Audio Engineering Associates MS380 TX:** Similar to MS38 plus high-pass filter, pad, gain trims, mic pre/matrix switch, and master level. See Figure 16–10.
- **Neumann Z 240:** Matrixing transformer pair.
- **Sanken MBB-II Battery Power Supply and Switchable Matrix Box for the Sanken CMS-7 Microphone:** MS or left-right output switch. See Figure 16–11.
- **Sanken M-5 Matrix Box for the Sanken CMS-2 Microphone:** Dual transformers.
- **Schoeps VMS 52U:** Variable MS matrix decodes in recording or in playback. Separate direct and matrixed outputs. Overload indicator, battery meter, 30 Hz fixed LF filter. Runs on 8 AA cell batteries or AC.

Figure 16–10 Audio Engineering Associates MS380 TX universal matrix box. (Courtesy Audio Engineering Associates.)

Figure 16–11 Sanken MBB-II matrix box. (Courtesy Pan Communications, Inc.)

- **Schoeps VMS 52U Lin:** Like the preceding but without a 30 Hz LF filter or external input to MS matrix. For stand-alone playback.
- **Schoeps VMS 02IB:** Like the VMS 52 but with Collette miniature stereo amplifiers and connectors. MS-XY selector switch for headphone monitoring. Ideal for portable DAT.

Company Addresses

AKG Acoustics US
1449 Donelson Pike
Nashville, TN 37217
615-399-2199
http://www.akg-acoustics.com

Audio Engineering Associates
1029 N. Allen Avenue
Pasadena, CA 91104
818-798-9127
stereoms@aol.com
http://members.aol.com/stereoms/
Telescoping mic stands also available.

Audio Technica US Incorporated
1221 Commerce Drive
Stow, OH 44224
330-686-2600
http://www.audio-technica.com

Beyerdynamic Incorporated
56 Central Avenue
Farmingdale, NY 11735
516-293-3200

Bruel & Kjaer Instruments, Incorporated
DK-2850 Naerum, Denmark
+45 42 800500
USA phone 800-332-2040 or 404-981-9311

Core Sound
http://www.panix.com/~moskowit/

Crown International Incorporated
1718 W. Mishawaka Road
Elkhart, IN 46517
219-294-8000
http://www.crownintl.com

HEAD Acoustics
(Aachen Head HMS II)
Sonic Perceptions, Incorporated, OS
28 Knight Street
Norwalk, CT 06851
203-838-2650
http://www.head-acoustics

Holophonics 3-D Sound Company
http://www.holophonics.com/index.htm
Josephson Engineering
3729 Corkerhill
San Jose, CA 95121
408-238-6062
http://www.josephson.com

Neumann USA
P.O. Box 987
Old Lyme, CT 06371
860-434-5220
http://www.neumannusa.com

Pearl Microphones AB
P.O. Box 98
S-265 21
Astorp, Sweden 46 42 588 10
http://www.pearl.se/

Sanken Microphones
213-845-1155

Sanken Microphones
c/o Turner Audio Sales
97 S. Main, Unit 10
Newtown, CT 06470
203-270-2933
http://www.turneraudio.com

Schoeps Microphones
c/o Posthorn Recordings
142 W. 26th Street
New York, NY 10001-6814
212-242-3737
http://www.posthorn.com

Schoeps Microphones
c/o Professional Sound Services Inc.
311 W. 43rd Street, #1100
New York, NY 10036-6413
800-883-1033
212-586-1033

Sennheiser Electronic Corporation
One Enterprise Drive
Old Lyme, CT 06371
860-434-9190
http://www.sennheiser.com

Shure Brothers Incorporated
222 Hartrey Avenue
Evanston, IL 60204
847-866-2200

Sony Professional Products
Sony Drive
Park Ridge, NJ 07656
201-939-1000
http://www.sony.com

Sonic Studios (DSM mics)
1311 Sunny Court
Sutherlin, OR 97479
541-459-8839
http://www.sonicstudios.com/index.htm#dsm

SoundField Research Limited
Charlotte Street Business Centre
Charlotte Street
Wakefield
West Yorkshire, United Kingdom
WF1 1UH
+44 (0) 1924 201089
http://www.proaudio.co.uk/sps422.htm

SoundField Microphones
c/o Transamerica Audio Group, Incorporated
2721 Calle Olivio
Thousand Oaks, CA 91360
805-241-4443

GLOSSARY

A-B—*See* Spaced-pair method.

ACCENT MICROPHONE—*See* Spot microphone.

ACCESS JACKS—Two jacks in a console input or output module that allow access to points in the signal path, usually for connecting a compressor or multitrack recorder. Plugging into the access jacks breaks the signal flow and allows you to insert a signal processor or recorder in series with the signal. Same as Insert jacks.

AMBIENCE—Room acoustics, early reflections, and reverberation. Also, the audible sense of a room or environment surrounding a recorded instrument.

AMBIENCE MICROPHONE—A microphone placed relatively far from its sound source to pick up ambience.

AMPLITUDE—Level, intensity, or magnitude. The amplitude of a sound wave or signal, as measured on a meter, is 0.707 times the peak amplitude. The peak amplitude is the voltage of the signal waveform peak.

ANALOG-TO-DIGITAL (A/D) CONVERTER—A circuit that converts an analog signal to a digital signal.

ARTIFICIAL HEAD—*See* Dummy head.

ASSIGN—To route or send an audio signal to one or more selected channels.

ATTENUATE—To reduce the level of a signal.

ATTENUATOR—In a mixer (or mixing console) input module, an adjustable resistive network that reduces the microphone signal level to prevent overloading of the input transformer and mic preamplifier.

AUXILIARY BUS (AUX-BUS)—*See* Effects bus.

AUXILIARY SEND (AUX-SEND)—*See* Effects send.

BAFFLED OMNI—A stereo miking arrangement that uses two ear-spaced omnidirectional microphones separated by a hard padded baffle.

BALANCE—The relative volume levels of tracks in a mix or instruments in a musical ensemble.

BALANCED LINE—A cable with two conductors surrounded by a shield, in which each conductor is at equal impedance to ground and opposite polarity; the signal flows through both conductors.

BINAURAL RECORDING—A two-channel recording made with an omnidirectional microphone mounted near each ear of a human or a dummy head for playback over headphones. The object is to duplicate the acoustic signal appearing at each ear.

BLUMLEIN ARRAY—A stereo microphone technique in which two coincident bidirectional microphones are angled 90° apart (45° to the left and right of center).

BOARD—*See* Mixing console.

BOUNDARY MICROPHONE—A microphone designed to be used on a boundary (a hard reflective surface). The microphone capsule is mounted very close to the boundary so that direct and reflected sounds arrive at the microphone diaphragm in phase (or nearly so) for all frequencies in the audible band.

BUS—A common connection of many different signals. The output of a mixer or submixer. A channel that feeds a tape track, signal processor, or power amplifier.

BUS IN—An input to a program bus, usually used for effects returns.

BUS MASTER—In the output section of a mixing console, a potentiometer (fader or volume control) that controls the output level of a bus.

BUS OUT—The output connector of a bus.

BUS TRIM—A control in the output section of a mixing console that provides variable gain control of a bus, used in addition to the bus master for fine adjustment.

BUZZ—An unwanted edgy tone that sometimes accompanies audio, containing high harmonics of 60 Hz (50 Hz in the U.K.).

CARDIOID MICROPHONE—A unidirectional microphone with side attenuation of 6 dB and maximum rejection of sound at the rear of the microphone (180° off axis). A microphone with a heart-shaped directional pattern.

CHANNEL—A single path of an audio signal. Usually, each channel contains a different signal.

CHANNEL ASSIGN—*See* Assign.

CLEAN—Free of noise, distortion, overhang, leakage; not muddy.

CLEAR—Easy to hear, easy to differentiate; reproduced with sufficient high frequencies.

CLOSELY SPACED METHOD—*See* Near-coincident method.

COINCIDENT-PAIR METHOD—A stereo microphone, or two separate microphones, placed so that the microphone diaphragms occupy approximately the same point in space. They are angled apart and mounted one directly above the other.

COMB-FILTER EFFECT—The frequency response caused by combining a sound with its delayed replica. The frequency response has a series of peaks and dips caused by phase interference. The peaks and dips resemble the teeth of a comb. This effect can occur with near-coincident and spaced-pair techniques when the left- and right-channel signals are combined to mono.

COMPRESSOR—A signal processor that reduces dynamic range by means of automatic volume control; an amplifier whose gain decreases as the input signal level increases above a preset point.

CONDENSER MICROPHONE—A microphone that works on the principle of variable capacitance to generate an electrical signal. The microphone diaphragm and an adjacent metallic disk (called a *backplate*) are charged to form two plates of a capacitor. Incoming sound waves vibrate the diaphragm, varying its spacing to the backplate, which varies the capacitance, which in turn varies the voltage between the diaphragm and backplate.

CONNECTOR—A device that makes electrical contact between a signal-carrying cable and an electronic device or between two cables. A device used to connect or hold together a cable and an electronic component so that a signal can flow from one to the other.

CONSOLE—*See* Mixing console.

CONTACT PICKUP—A transducer that contacts a musical instrument and converts its mechanical vibrations into a corresponding electrical signal.

CROSSTALK—The unwanted transfer of a signal from one channel to another. Head-related crosstalk is the right-speaker signal that reaches the left ear and the left-speaker signal that reaches the right ear. In the transaural stereo system, this acoustic crosstalk is canceled by processing the stereo signal with electronic crosstalk that is the inverse of the acoustic crosstalk.

DAT (R-DAT)—A digital audiotape recorder that uses a rotating head to record digital audio on tape.

DAW—Acronym for digital audio workstation; *See* Digital audio workstation.

dB—Abbreviation for decibel; *See* Decibel.

dBA—Decibels, A weighted; *See* Weighted.

dBm—Decibels relative to 1 milliwatt.

dBu—Decibels relative to 0.775 volt.

dBV—Decibels relative to 1 volt.

DEAD—Having very little or no reverberation.

DECAY—The portion of the envelope of a note in which the envelope goes from a maximum level to some midrange level. Also, the decline in level of reverberation over time.

DECAY TIME—*See* Reverberation time.

DECIBEL—The unit of measurement of audio level: 10 times the logarithm of the ratio of two power levels; 20 times the logarithm of the ratio of two voltages.

DELAY—The time interval between a signal and its repetition. A digital delay or a delay line is a signal processor that delays a signal for a short time.

DEMAGNETIZER (DEGAUSSER)—An electromagnet with a probe tip that is touched to elements of the tape path (such as tape heads and tape guides) to remove residual magnetism.

DEPTH—The audible sense of nearness and farness of various instruments. Instruments recorded with a high ratio of direct-to-reverberant sound are perceived as close; instruments recorded with a low ratio of direct-to-reverberant sound are perceived as distant.

DESIGN CENTER—The portion of fader travel (usually shaded), about 10–15 dB from the top, in which console gain is distributed for optimum headroom and signal-to-noise ratio. During normal operation, each fader in use should be placed at or near design center.

DESIGNATION STRIP—A strip of paper taped near console faders to designate the instrument that each fader controls.

DESK—The British term for mixing console.

DESTRUCTIVE EDITING—In a digital audio workstation, editing that rewrites the data on disk. A destructive edit cannot be undone.

DI—Acronym for direct injection, recording with a direct box.

DIFFUSE FIELD—A sound field in which the sounds arrive randomly from all directions, such as the reverberant field in a concert hall. Diffuse-field equalization is equalization applied to a dummy head so that it has a net flat response in a diffuse sound field.

DIGITAL AUDIO—Encoding an analog audio signal in the form of binary digits (ones and zeros).

DIGITAL AUDIO WORKSTATION (DAW)—A computer, sound card, and editing software that allows recording, editing, and mixing audio programs entirely in digital form. Stand-alone DAWs include real mixer controls; computer DAWs have virtual controls on screen.

DIGITAL RECORDING—A recording system in which the audio signal is stored in the form of binary digits.

DIGITAL-TO-ANALOG CONVERTER—A circuit that converts a digital audio signal into an analog audio signal.

DIRECT BOX—A device used for connecting an amplified instrument directly to a mixer mic input. The direct box converts a high-impedance, unbalanced audio signal into a low-impedance, balanced audio signal.

DIRECT INJECTION (DI)—Recording with a direct box.

DIRECTIONAL MICROPHONE—A microphone that has different sensitivity in different directions. A unidirectional or bidirectional microphone.

DIRECT OUTPUT, DIRECT OUT—An output connector following a mic preamplifier, fader, and equalizer, used to feed the signal of one instrument to one track of a tape recorder.

DIRECT SOUND—Sound traveling directly from the sound source to the microphone (or to the listener) without reflections.

DISTORTION—An unwanted change in the audio waveform, causing a raspy or gritty sound quality. The appearance of frequencies in a device's output signal that were not in the input signal. Distortion is caused by recording at too high a level, improper mixer settings, components failing, or vacuum tubes distorting. (Distortion can be desirable; for an electric guitar, for example.)

DROPOUT—During playback of a tape recording, a momentary loss of signal caused by separation of the tape from the playback head due to dust, tape-oxide irregularity, and so forth.

DRY—Having no echo or reverberation. Refers to a close-sounding signal that is not yet processed by a reverberation or delay device.

DSD—Acronym for Direct Stream Digital, a Sony trademark for 1-bit encoding of digital signals used in the company's Super Audio CD format.

DSP—Acronym for digital signal processing, modifying a signal in digital form.

DUMMY HEAD—A modeled head with microphones in the ears, used for binaural recording; same as artificial head.

DVD—A digital versatile disc. A storage medium the size of a compact disc, which holds much more data. The DVD stores video, audio, or computer data.

DYNAMIC MICROPHONE—A microphone that generates electricity when sound waves cause a conductor to vibrate in a stationary magnetic field. The two types of dynamic microphone are moving coil and ribbon. A moving-coil microphone usually is called a *dynamic microphone.*

DYNAMIC RANGE—The range of volume levels in a program from softest to loudest.

EARTH GROUND—A connection to moist dirt (the ground we walk on). This connection usually is done via a long copper rod or an all-metal cold-water pipe.

ECHO—A delayed repetition of a signal or sound. A sound delayed 50 milliseconds or more, combined with the original sound.

EDIT DECISION LIST (EDL)—A list of program events in order, plus their starting and ending times.

EDITING—Cutting and rejoining magnetic tape to delete unwanted material, to insert leader tape, or to rearrange recorded material into the desired sequence. Also, the same actions performed with a digital audio workstation, hard disk recorder, or MiniDisc recorder-mixer without cutting any tape.

EFFECTS—Interesting sound phenomena created by signal processors, such as reverberation, echo, flanging, doubling, compression, or chorus.

EFFECTS BUS—The bus that feeds effects devices (signal processors).

EFFECTS LOOP—A set of connectors in a mixer for connecting an external effects unit, such as a reverb or delay device. The effects loop includes a send section and a receive section. *See* Effects send, Effects return.

EFFECTS MIXER—A submixer in a mixing console that combines signals from effect sends and then feeds the mixed signal to the input of an effects device, such as a reverberation unit.

EFFECTS RETURN (EFFECTS RECEIVE)—In the output section of a mixing console, a control that adjusts the amount of signal received from an effects unit. Also, the connectors in a mixer to which you connect the effects-unit output signal. It might be labeled *bus in* instead. The effects-return signal is mixed with the program bus signal.

EFFECTS SEND—In an input module of a mixing console, a control that adjusts the amount of signal sent to an effects device, such as a reverberation or delay unit. Also, the connector in a mixer to which you connect the input of an effects unit. The effects-send control normally adjusts the amount of reverberation or echo heard on each instrument.

ELECTRET-CONDENSER MICROPHONE—A condenser microphone in which the electrostatic field of the capacitor is generated by an electret, a material that permanently stores an electrostatic charge.

ELECTROSTATIC FIELD—The force field between two conductors charged with static electricity.

ELECTROSTATIC INTERFERENCE—The unwanted presence of an electrostatic hum field in signal conductors.

ELEVATION—An image displacement in height above the speaker plane.

END ADDRESSED—Referring to a microphone whose main axis of pickup is perpendicular to the front of the microphone. You aim the front of the mic at the sound source. *See* Side addressed.

ENVELOPE—The rise and fall in volume of one note. The envelope connects successive peaks of the waves constituting a note. Each harmonic in the note might have a different envelope.

EQUALIZATION (EQ)—The adjustment of frequency response to alter the tonal balance or to attenuate unwanted frequencies.

EQUALIZER—A circuit (usually in each input module of a mixing console or in a separate unit) that alters the frequency spectrum of a signal passed through it.

FADE OUT—Gradually reduce the volume of the last several seconds of a recorded song, from full level down to silence, by slowly pulling down the master fader.

FADER—A linear or sliding potentiometer (volume control), used to adjust signal level.

FAULKNER METHOD—Named after Tony Faulkner, a stereo microphone technique using two bidirectional microphones aiming at the sound source and spaced about 8 inches apart.

FEED—1. To send an audio signal to some device or system. 2. An output signal sent to some device or system.

FEEDBACK—1. The return of some portion of an output signal to the system's input. 2. The squealing sound you hear when a PA system microphone picks up its own amplified signal through a loudspeaker.

FILTER—1. A circuit that sharply attenuates frequencies above or below a certain frequency. Used to reduce noise and leakage above or below the frequency range of an instrument or voice. 2. A MIDI filter removes selected note parameters.

FLOAT—Disconnect from ground.

FOCUS—The degree of fusion, compactness, or positional definition of a sonic image.

FREE FIELD—The sound field coming directly from the sound source with no reflections; the sound field in an anechoic chamber. Free-field equalization is equalization applied to a dummy head to make it have a net flat response in a free field.

FREQUENCY—The number of cycles per second of a sound wave or an audio signal, measured in hertz (Hz). A low frequency (for example, 100 Hz) has a low pitch; a high frequency (for example, 10,000 Hz) has a high pitch.

FREQUENCY RESPONSE—1. The range of frequencies that an audio device will reproduce at an equal level (within a tolerance, such as ±3 dB). 2. The range of frequencies that a device (mic, human ear, etc.) can detect.

FUNDAMENTAL—The lowest frequency in a complex wave.

FUSION—The formation of a single image by two or more sound sources, such as loudspeakers.

GAIN—Amplification. The ratio, expressed in decibels, between the output voltage and the input voltage or between the output power and the input power.

GENERATION—A copy of a tape. A copy of the original master recording is a first generation tape. A copy made from the first generation tape is a second generation tape, and so on.

GENERATION LOSS—The degradation of signal quality (the increase in noise, distortion, and phase shift) that occurs with each successive generation of an analog tape recording or with each A/D, D/A conversion.

GRAPHIC EQUALIZER—An equalizer with a horizontal row of faders; the fader-knob positions indicate graphically the frequency response of the equalizer. Usually used to equalize monitor speakers for the room they are in; sometimes used for complex EQ of a track.

GROUND—The zero-signal reference point for a system of audio components.

GROUND BUS—A common connection to which equipment is grounded, usually a heavy copper plate.

GROUNDING—Connecting pieces of electronic equipment to ground. Proper grounding ensures that there is no voltage difference between equipment chassis. An electrostatic shield needs to be grounded to be effective.

GROUND LOOP—1. A loop or circuit formed of ground leads. 2. The loop formed when components are connected together via two ground paths: the connecting-cable shield and the power ground. Ground loops cause hum and should be avoided.

GROUP—*See* Submix.

HARD DISK—A random-access storage medium for computer data. A hard disk drive contains a stack of magnetically coated hard disks that are read and written to by an electromagnetic head.

HARD DISK RECORDER—A device dedicated to recording digital audio on a hard disk drive. A hard disk recorder-mixer includes a built-in mixer.

HARMONIC—In a complex wave, an overtone whose frequency is a whole-number multiple of the fundamental frequency.

HEADROOM—The safety margin, measured in decibels, between the signal level and the maximum undistorted signal level. In an analog tape recorder, the dB difference between standard operating level (corresponding to a 0 VU reading) and the level causing 3 percent total harmonic distortion. High-frequency headroom increases with analog tape speed.

HERTZ (Hz)—Cycles per second, the unit of measurement of frequency.

HIGH-PASS FILTER—A filter that passes frequencies above a certain frequency and attenuates frequencies below that same frequency. A low-cut filter.

HISS—A noise signal containing all frequencies, but with greater energy at higher octaves. Hiss sounds like wind blowing through trees. It usually is caused by random signals generated by microphones, electronics, and magnetic tape.

HOT—1. A high recording level causing slight distortion, maybe used for special effect. 2. A condition in which a chassis or circuit carries a potentially dangerous voltage. 3. Refers to the conductor in a microphone cable that has a positive voltage on it at the instant that sound pressure moves the diaphragm inward.

HUM—An unwanted low-pitched tone (60 Hz and its harmonics) heard in the monitors. The sound of interference generated in audio circuits and cables by AC power wiring. Hum pickup is caused by such things as faulty grounding, poor shielding, and ground loops.

HYPERCARDIOID MICROPHONE—A directional microphone with a polar pattern that has 12 dB attenuation at the sides, 6 dB attenuation at the rear, and two nulls of maximum rejection at 110° off axis.

IMAGE—An illusory sound source located somewhere around the listener. An image is generated by two or more loudspeakers. In a typical stereo system, images are located between the two stereo speakers.

IMAGING—The ability of a microphone array or a speaker pair to form easily localizable images.

IMPEDANCE—The opposition of a circuit to the flow of alternating current. Impedance is the complex sum of resistance and reactance, represented mathematically by Z.

INPUT—The connection going into an audio device. In a mixer or mixing console, a connector for a microphone, line-level device, or other signal source.

INPUT ATTENUATOR—*See* Attenuator.

INPUT MODULE—In a mixing console, the set of controls affecting a single input signal. An input module usually includes an attenuator (trim), fader, equalizer, aux sends, and channel-assign controls.

INPUT SECTION—The row of input modules in a mixing console.

INPUT/OUTPUT (I/O) CONSOLE (IN-LINE CONSOLE)—A mixing console arranged so that input and output sections are aligned vertically. Each module (other than the monitor section) contains one input channel and one output channel.

INSERT JACKS—*See* Access jacks.

INTENSITY STEREO (XY STEREO)—A method of forming stereo images by intensity or amplitude differences between channels; *See* Coincident-pair method.

ITE/PAR—Acronym for in the ear/pinna acoustic response. A stereo recording system developed by Don and Carolyn Davis of Synergetic Audio Concepts, it uses two probe microphones in the ear canals, near the ear drums of a human listener. Playback is over two speakers up front and two to the sides of the listener.

JACK—A female or receptacle-type connector for audio signals into which a plug is inserted.

JECKLIN DISK—Named after its inventor, a stereo microphone array using two omnidirectional microphones spaced $6\frac{1}{2}$ inches apart and separated by a disk or baffle $11\frac{7}{8}$ inches in diameter, covered with sound absorbent material; also known as the *OSS System.*

KILO—A prefix meaning 1000, abbreviated k.

LEAKAGE—The overlap of an instrument's sound into another instrument's microphone. Also called *bleed* or *spill.*

LEVEL—The degree of intensity of an audio signal: the voltage, power, or sound pressure level. Originally, the power in watts.

LEVEL SETTING—In a recording system, the process of adjusting the input-signal level to obtain the maximum level on the recording media with no distortion. A VU meter or other indicator shows the recording level.

LIMITER—A signal processor whose output is constant above a preset input level. A compressor with a compression ratio of 10:1 or greater, with the threshold set just below the point of distortion of the following device. Used to prevent distortion of attack transients or peaks.

LINE LEVEL—In balanced professional recording equipment, a signal whose level is approximately 1.23 volts (+4 dBm). In unbalanced equipment (most home hi-fi or semipro recording equipment), a signal whose level is approximately 0.316 volt (-10 dBV).

LIVE—1. Having audible reverberation. 2. Occurring in real time, in person.

LIVE RECORDING—A recording made at a concert. Also, a recording made of a musical ensemble playing all at once, rather than overdubbing.

LOCALIZATION—The ability of the human hearing system to tell the direction of a real or illusory sound source.

LOCALIZATION ACCURACY—The accuracy with which a stereo microphone array translates the location of real sound sources into image locations. If localization is accurate, instruments at the side of the musical ensemble are reproduced from the left or right speaker; instruments halfway off center are reproduced halfway between the center of the speaker pair and one speaker, and so on.

LOCATION—The angular position of an image relative to a point straight ahead of a listener or its position relative to the loudspeakers.

M—Abbreviation for mega, or 1 million (as in megabytes).

MASTER FADER—A volume control that affects the level of all program buses simultaneously. It is the last stage of gain adjustment before the two-track recorder.

MASTER TAPE—A completed tape used to generate tape copies or compact discs.

MDM—Acronym for modular digital multitrack.

METER—A device that indicates voltage, resistance, current, or signal level.

MIC—Short for microphone; *See* Microphone.

MIC LEVEL—The level or voltage of a signal produced by a microphone, typically 2 millivolts.

MIC PREAMP—*See* Preamplifier.

MICROPHONE—A transducer or device that converts an acoustical signal (sound) into a corresponding electrical signal.

MICROPHONE TECHNIQUES—The selection and placement of microphones to pick up sound sources.

MID-SIDE METHOD—A coincident-pair stereo microphone technique using a forward-facing unidirectional, omnidirectional, or bidirectional mic and a side-facing bidirectional mic. The microphone signals are summed and differenced to produce right- and left-channel signals.

MILLI—A prefix meaning 1000th, abbreviated m.

MIKE—To pick up with a microphone.

MIX—1. To combine two or more different signals into a common signal. 2. A control on a delay unit that varies the ratio between the dry signal and the delayed signal.

MIXDOWN—The process of playing recorded tape tracks through a mixing console and mixing them to two stereo channels for recording on a two-track tape recorder.

MIXER—A device that mixes or combines audio signals and controls the relative levels of the signals.

MIXING CONSOLE—A large mixer with additional functions such as equalization or tone control, pan pots, monitoring controls, solo functions, channel assigns, and control of signals sent to external signal processors.

MODULAR DIGITAL MULTITRACK (MDM)—A multitrack tape recorder that records eight tracks digitally on a videocassette. Several eight-track modules can be connected to add more tracks in sync. Two examples of MDMs are the Alesis ADAT-XT and TASCAM DA-38.

MONAURAL—Refers to listening with one ear; often incorrectly used to mean monophonic.

MONITOR—To listen to an audio signal with headphones or loudspeakers. Also, a loudspeaker in a control room, or headphones, used for judging sound quality. Also, a video display screen used with a computer or video camera.

MONO, MONOPHONIC—1. Referring to a single channel of audio. A monophonic program can be played over one or more loudspeakers or one or more headphones. 2. Describing a synthesizer that plays only one note at a time (not chords).

MONO COMPATIBLE—A characteristic of a stereo program, in which the program channels can be combined to a mono program without altering the frequency response or balance. A mono-compatible stereo program has the same frequency response in stereo or mono because there is no delay or phase shift between channels to cause phase interference.

MOVING-COIL MICROPHONE—A dynamic microphone in which the conductor is a coil of wire moving in a fixed magnetic field. The coil is attached to a diaphragm that vibrates when struck with sound waves. Usually called a *dynamic microphone*.

MS RECORDING—*See* Mid-side method.

MUDDY—Unclear sounding; having excessive leakage, reverberation, or overhang.

MULTIPLE-D MICROPHONE—A directional microphone that has multiple sound-path lengths between its front and rear sound entries. This type of microphone has a minimal proximity effect.

MULTITRACK—Refers to a recorder or tape-recorder head that has more than two tracks.

MUTE—To turn off an input signal on a mixing console by disconnecting the input-module output from channel assign and direct out. During mixdown, the mute function is used to reduce tape noise during silent portions of tracks or to turn off unused performances. During recording, mute is used to turn off mic signals.

NEAR-COINCIDENT-PAIR METHOD—A stereo microphone technique in which two directional microphones are angled apart symmetrically on either side of center and spaced a few inches apart horizontally.

NEAR-FIELD™ MONITORING—A monitor speaker arrangement in which the speakers are placed very near the listener (usually just behind the mixing console) to reduce the audibility of control-room acoustics.

NOISE—Unwanted sound, such as hiss from electronics or tape. An audio signal with an irregular, nonperiodic waveform.

NONDESTRUCTIVE EDITING—In a digital audio workstation, editing done by changing pointers (location markers) to information on the hard disk. A nondestructive edit can be undone.

NONLINEAR—1. Refers to a storage medium in which any data point can be accessed or read almost instantly. Examples are a hard disk, compact disc, and MiniDisc. *See* Random access. 2. Refers to an audio device that is distorting the signal.

NOS SYSTEM—A near-coincident stereo microphone technique in which two cardioid microphones are angled 90° and spaced 30 cm apart horizontally.

OFF AXIS—Not directly in front of a microphone or loudspeaker.

OFF-AXIS COLORATION—In most microphones, the deviation from the on-axis frequency response that occurs at angles off the axis of the microphone. The coloration of sound (alteration of tone quality) for sounds arriving off axis to the microphone.

OMNIDIRECTIONAL MICROPHONE—A microphone that is equally sensitive to sounds arriving from all directions.

ON-LOCATION RECORDING—A recording made outside the studio, in a room or hall where the music usually is performed or practiced.

ORTF—Named after the French broadcasting network (Office de Radiodiffusion Television Française), a near-coincident stereo mic technique that uses two cardioid mics angled 110° and spaced 17 cm (6.7 inches) apart horizontally.

OSS SYSTEM—Acronym for optimal stereo signal system. *See* Jecklin disk.

OUTPUT—A connector in an audio device from which the signal comes and feeds successive devices.

OVERDUB—Record a new musical part on an unused track in synchronization with previously recorded tracks.

OVERLOAD—The distortion that occurs when an applied signal exceeds a system's maximum input level.

PAD—*See* Attenuator.

PAN POT—Short for panoramic potentiometer. In each input module in a mixing console, this control divides a signal between two channels in an adjustable ratio. By doing so, a pan pot controls the location of a sonic image between a stereo pair of loudspeakers.

PARABOLIC MICROPHONE—A highly directional microphone made of a parabola-shaped sound reflector that focuses sound into the microphone capsule.

PATCH—1. To connect one piece of audio equipment to another with a cable. 2. A setting of synthesizer parameters to achieve a sound with a certain timbre.

PATCH BAY (PATCH PANEL)—An array of connectors, usually in a rack, to which equipment input and output lines are wired. A patch bay makes it easy to interconnect various pieces of equipment in a central, accessible location.

PATCH CORD—A short length of cable with a coaxial plug on each end, used for signal routing in a patch bay.

PERSPECTIVE—In the reproduction of a recording, the audible sense of distance to the musical ensemble, the point of view. A close perspective has a high ratio of direct sound to reverberant sound; a distant perspective has a low ratio of direct sound to reverberant sound.

PFL—Acronym for prefader listen. *See also* Solo.

PHANTOM IMAGE—*See* Image.

PHANTOM POWER—DC voltage (usually 12 to 48 volts) applied to microphone signal conductors to power condenser microphones.

PHASE—The degree of progression in the cycle of a wave, where one complete cycle is 360°.

PHASE CANCELLATION, PHASE INTERFERENCE—The cancellation of certain frequency components of a signal that occurs when the signal is combined with its delayed replica. At certain frequencies, the direct and delayed signals are of equal level and opposite polarity (180° out of phase), and when combined, they cancel out. The result is a comb-filter frequency response having a periodic series of peaks and dips. Phase interference can occur between the signals of two microphones picking up the same source at different distances or at a microphone picking up both a direct sound and its reflection from a nearby surface.

PHASE SHIFT—The difference in degrees of phase angle between corresponding points on two waves. If one wave is delayed with respect to another, there is a phase shift between them of $2\pi FT$, where $\pi = 3.14$, $F =$ frequency in Hz, and $T =$ delay in seconds.

PHONE PLUG—A cylindrical, coaxial plug (usually $\frac{1}{4}$-inch diameter). An unbalanced phone plug has a tip for the hot signal and a sleeve for the shield or ground. A balanced phone plug has a tip for the hot signal, a ring for the return signal, and a sleeve for the shield or ground.

PHONO PLUG—A coaxial plug with a central pin for the hot signal and a ring of pressure-fit tabs for the shield or ground. Also called an *RCA plug*.

PICKUP—A piezoelectric transducer that converts mechanical vibrations to an electrical signal. Used in acoustic guitars, acoustic basses, and fiddles. Also, a magnetic transducer in an electric guitar that converts string vibration to a corresponding electrical signal.

PINNAE—The outer ears. Reflections from folds of skin in the pinnae aid in localizing sounds.

PLAYLIST—*See* Edit decision list.

PLUG—A male connector that inserts into a jack.

PLUG-INS—Software effects that you install in your computer. The plug-in software becomes part of another program you are using, such as a digital editing program.

POLAR PATTERN—The directional pickup pattern of a microphone. A plot of microphone sensitivity plotted versus the angle of sound incidence. Examples of polar patterns are omnidirectional, bidirectional, and unidirectional. Subsets of unidirectional are cardioid, supercardioid, and hypercardioid.

POLARITY—Refers to the positive or negative direction of an electrical, acoustical, or magnetic force. Two identical signals in opposite polarity are 180° out of phase with each other at all frequencies.

POP—1. A thump or little explosion sound heard in a vocalist's microphone signal. Pops occur when the user says words with *p*, *t*, or *b* so that a turbulent puff of air is forced from the mouth and strikes the microphone diaphragm. 2. A noise heard when a mic is plugged into a monitored channel or when a switch is flipped.

POP FILTER—A screen placed on a microphone grille that attenuates or filters out pop disturbances before they strike the microphone diaphragm. Usually made of open-cell plastic foam or silk, a pop filter reduces pop and wind noise.

POWER AMPLIFIER—An electronic device that amplifies or increases the power level fed into it to a level sufficient to drive a loudspeaker.

POWER GROUND (SAFETY GROUND)—A connection to the power company's earth ground through the U-shaped hole in a power outlet. In the power cable of an electronic component with a three-prong plug, the U-shaped prong is wired to the component's chassis. This wire conducts electricity to power ground if the chassis becomes electrically hot, preventing shocks.

PREAMPLIFIER (PREAMP)—In an audio system, the first stage of amplification that boosts a mic-level signal to line level. A preamp is a standalone device or a circuit in a mixer.

PREFADER-POSTFADER SWITCH—A switch that selects a signal either ahead of (prefader) or following (postfader) the fader. The level of a prefader signal is independent of the fader position; the level of a postfader signal follows the fader position.

PREPRODUCTION—Planning in advance what will be done at a recording session, in terms of track assignments, overdubbing, studio layout, and microphone selection.

PRESSURE ZONE MICROPHONE—A boundary microphone constructed with the microphone diaphragm parallel to and facing a reflective surface.

PRODUCTION—1. A recording that is enhanced by special effects. 2. The supervision of a recording session to create a satisfactory recording. This involves getting musicians together for the session, making musical suggestions to the musicians to enhance their performance, and making suggestions to the engineer for sound balance and effects.

PROGRAM BUS—A bus or output that feeds an audio program to a recorder track.

PROGRAM MIXER—In a mixing console, a mixer formed of input-module output, combining amplifiers, and program buses.

PROXIMITY EFFECT—The bass boost that occurs with a single-D directional microphone when it is placed a few inches from a sound source. The closer the microphone, the greater is the low-frequency boost due to the proximity effect.

RACK—A 19-inch-wide wooden or metal cabinet used to hold audio equipment.

RADIO-FREQUENCY INTERFERENCE (RFI)—Radio-frequency electromagnetic waves induced in audio cables or equipment, causing various noises in the audio signal.

RANDOM ACCESS—Refers to a storage medium in which any data point can be accessed or read almost instantly. Examples are a hard disk, compact disk, and MiniDisc.

R-DAT—*See* DAT.

RECORDER-MIXER—A combination multitrack recorder and mixer in one chassis.

REFLECTED SOUND—Sound waves that reach the listener after being reflected from one or more surfaces.

REGION—In a digital audio editing program, a defined segment of the audio program.

REMOTE RECORDING—*See* On-location recording.

REMOVABLE HARD DRIVE—A hard disk drive that can be removed and replaced with another, used in a digital audio workstation to store a long program temporarily.

REVERBERATION—Natural reverberation in a room is a series of multiple sound reflections that makes the original sound persist and gradually die away or decay. These reflections tell the ear that you're listening in a large or hard-surfaced room. For example, reverberation is the sound you hear just after you shout in an empty gymnasium. A reverb effect simulates the sound of a room—club, auditorium, or concert hall—by generating random multiple echoes that are too numerous and rapid for the ear to resolve. The timing of the echoes is random, and the echoes increase in number with time as they decay. An echo is a discrete repetition of a sound; reverberation is a continuous fade-out of sound.

REVERBERATION TIME (RT60)—The time it takes for reverberation to decay to 60 dB below the original steady-state level.

RFI—*See* Radio frequency interference.

RIBBON MICROPHONE—A dynamic microphone in which the conductor is a long metallic diaphragm (ribbon) suspended in a magnetic field. Usually a ribbon microphone has a bidirectional (figure-eight) polar pattern and can be used for the Blumlein method of stereo recording.

SAMPLING—Recording a short sound event into computer memory. The audio signal is converted into digital data representing the signal waveform, and the data are stored in memory chips or on disk for later playback.

SASS—The Stereo Ambient Sampling System™, a stereo microphone using two boundary microphones, each on a 5-inch-square panel, angled apart and ear spaced, with a baffle between the microphones.

SEMI-COINCIDENT METHOD—*See* Near-coincident-pair method.

SENSITIVITY—1. The output of a microphone in volts for a given input in sound pressure level. 2. The sound pressure level a loudspeaker produces at one meter when driven with 1 watt of pink noise. *See also* Sound pressure level.

SHIELD—A conductive enclosure (usually metallic) around one or more signal conductors, used to keep out electrostatic fields that cause hum or buzz.

SHOCK MOUNT—A suspension system that mechanically isolates a microphone from its stand or boom, preventing the transfer of mechanical vibrations.

SHOTGUN MICROPHONE (LINE MICROPHONE)—A highly directional microphone made of a slotted "line interference" tube mounted in front of a hypercardioid microphone capsule.

SHUFFLING—*See* Spatial equalization.

SIBILANCE—In a speech recording, excessive frequency components in the 5–10 kHz range, which are heard as an overemphasis of *s* and *sh* sounds.

SIDE ADDRESSED—Refers to a microphone whose main axis of pickup is perpendicular to the side of the microphone. You aim the side of the mic at the sound source. *See also* End addressed.

SIGNAL—A varying electrical voltage that represents information, such as a sound.

SIGNAL PATH—The path a signal travels from input to output in a piece of audio equipment.

SIGNAL PROCESSOR—A device used to alter a signal in a controlled way.

SIGNAL-TO-NOISE (S/N) RATIO—The ratio in decibels between signal voltage and noise voltage. An audio component with a high S/N has little background noise accompanying the signal; a component with a low S/N is noisy.

SINGLE ENDED—1. An unbalanced line. 2. A single-ended noise reduction system works only during tape playback (unlike Dolby or dbx, which work during both recording and playback).

SINGLE-D MICROPHONE—A directional microphone having a single distance between its front and rear sound entries. Such a microphone has a proximity effect.

SIZE—*See* Focus.

SMPTE TIME CODE—A modulated 1200 Hz square-wave signal used to synchronize two or more tape transports or other multitrack recorders. SMPTE is an acronym for the Society of Motion Picture and Television Engineers, which developed the time code.

SNAKE—A multipair or multichannel mic cable. Also, a multipair mic cable attached to a connector junction box.

SOLO—On an input module in a mixing console, a switch that lets you monitor that particular input signal by itself. The switch routes only that input signal to the monitor system.

SOUND CARD—A circuit card that plugs into a computer and converts an audio signal into computer data for storage in memory or on hard disk. The sound card also converts computer data into an audio signal.

SOUND PRESSURE LEVEL (SPL)—The acoustic pressure of a sound wave, measured in decibels above the threshold of hearing: dB SPL = 20 log (P/P_{ref}), where $P_{ref} = 0.0002$ dyne/cm^2.

SPACED-PAIR METHOD—A stereo microphone technique using two identical microphones spaced several feet apart horizontally, usually aiming straight ahead toward the sound source.

SPATIAL EQUALIZATION—A low-frequency shelving boost in the L – R (difference) signal of a stereo program, and a complementary shelving cut in the L + R (sum) signal, to align the locations of the low- and high-frequency components of images and to increase spaciousness or stereo separation.

SPATIAL PROCESSOR—A signal processor that allows images to be placed beyond the limits of a stereo pair of speakers, even behind the listener or toward the sides.

SPECTRUM—The output versus frequency of a sound source, including the fundamental frequency and overtones.

SPL—*See* Sound pressure level.

SPLITTER—A transformer or circuit used to divide a microphone signal into two or more identical signals to feed different sound systems.

SPOT MICROPHONE—In classical music recording, a close-up microphone that is mixed with more-distant microphones to add presence or to improve the balance.

STAGE WIDTH—*See* Stereo spread.

STEREO, STEREOPHONIC—An audio recording and reproduction system with correlated information between two channels (usually discrete channels) and meant to be heard over two or more loudspeakers to give the illusion of sound-source localization and depth. *Stereo* means "solid" or three dimensional.

STEREO BAR, STEREO MICROPHONE ADAPTER—A microphone stand adapter that mounts two microphones on a single stand for convenient stereo miking.

STEREO IMAGING—The ability of a stereo recording or reproduction system to form clearly defined audio images at various locations between a stereo pair of loudspeakers.

STEREO MICROPHONE—A microphone containing two mic capsules in a single housing for convenient stereo recording. The capsules usually are coincident.

STEREO SPREAD—The reproduced stage width. The distance between the reproduced images of the left and right sides of a musical ensemble.

SUBMASTER—1. A master volume control for an output bus. 2. A recorded tape that is used to form a master tape.

SUBMIX—A small preset mix within a larger mix, such as a drum mix, keyboard mix, or vocal mix. Also a cue mix, monitor mix, or effects mix.

SUBMIXER—A smaller mixer within a mixing console (or standing alone) used to set up a submix, a cue mix, an effects mix, or a monitor mix.

SUPER AUDIO CD—Proposed by Sony, a compact-disc format with two layers. One layer contains a two-channel DSD program followed by the same program in six channels for surround. The other layer contains a two-channel 16 bit/44.1 K linear PCM program for compatibility with existing compact-disk players.

SUPERCARDIOID MICROPHONE—A unidirectional microphone that attenuates side-arriving sounds by 8.7 dB, attenuates rear-arriving sounds by 11.4 dB, and has two nulls of maximum sound rejection at 125° off axis.

SURROUND SOUND—A multichannel recording and reproduction system that plays sound all around the listener. The 5.1 surround-sound system uses the following speakers: front left, center, front right, left surround, right surround, and subwoofer.

SYNC, SYNCHRONIZATION—Alignment of two separate audio programs in time, and maintenance of that alignment as the programs play.

SYNC, SYNCHRONOUS RECORDING—Using a record head temporarily as a playback head during an overdub session to keep the overdubbed parts in synchronization with the recorded tracks.

SYNC TONE—*See* Tape sync.

SYNC TRACK—A track of a multitrack recorder that is reserved for recording an FSK sync tone or SMPTE time code. This allows audio tracks to synchronize with virtual tracks recorded with a sequencer. A sync track also can synchronize two audiotape machines or an audio recorder and a video recorder and can be used for console automation.

TAPE RECORDER—A device that converts an electrical audio signal into a magnetic audio signal on magnetic tape and vice versa. A tape recorder includes electronics, heads, and a transport to move the tape.

TAPE SYNC—A frequency-modulated signal recorded on a tape track, used to synchronize a tape recorder to a sequencer. Tape sync also permits the synchronized transfer of sequences to tape. *See also* Sync track.

THREE-PIN CONNECTOR—A three-pin professional audio connector used for balanced signals. Pin 1 is soldered to the cable shield, pin 2 is soldered to the signal hot lead, and pin 3 is soldered to the signal return lead. *See also* XLR-type connector.

THREE-TO-ONE RULE—A rule in microphone applications. When multiple mics are mixed to the same channel, the distance between mics should be at least three times the distance from each mic to its sound source. This prevents audible phase interference.

TIE—Connect electrically; for example, by soldering a wire between two points in a circuit.

TIGHT—1. Having very little leakage or room reflections in the sound pickup. 2. Refers to well-synchronized playing of musical instruments. 3. Having a well-damped, rapid decay.

TIMBRE—The subjective impression of spectrum and envelope. The quality of a sound that allows us to differentiate it from other sounds. For example, if you hear a trumpet, a piano, and a drum, each has a different timbre or tone quality that identifies it as that particular instrument.

TIME CODE—A modulated 1200 Hz square-wave signal used to synchronize two or more tape or disc transports. *Se also* Sync track, SMPTE.

TONAL BALANCE—The balance or volume relationships among different regions of the frequency spectrum, such as bass, midbass, midrange, upper midrange, and highs.

TRACK—A path on magnetic tape containing a single channel of audio. A group of bytes in a digital signal (on tape, hard disk, compact disk, or in a data stream) that represents a single channel of audio or MIDI. Usually one track contains a performance of one musical instrument.

TRANSAURAL STEREO—A method of stereo recording for surround sound. During recording, the signals from a dummy head are processed for playback over loudspeakers, so that acoustic crosstalk around the head is canceled. This crosstalk is the signal from the right speaker that reaches the left ear and the signal from the left speaker that reaches the right ear. The net effect is to enable the listener to hear, over loudspeakers, what the dummy head heard in the original environment.

TRANSDUCER—A device that converts energy from one form to another, such as a microphone or loudspeaker.

TRANSFORMER—An electronic component made of two magnetically coupled coils of wire. The input signal is transferred magnetically to the output, with no direct connection between input and output.

TRANSIENT—A rapidly changing signal with a fast attack and a short decay, such as a drum beat.

TRIM—1. In a mixing console, a control for fine adjustment of level, as in a bus trim control. 2. In a mixing console, a control that adjusts the gain of a mic preamp to accommodate various signal levels.

TUBE—A vacuum tube, an amplifying component made of electrodes in an evacuated glass tube. Tube sound is characterized as being "warmer" than solid-state or transistor sound.

UNBALANCED LINE—An audio cable having one conductor surrounded by a shield that carries the return signal. The shield is at ground potential.

UNIDIRECTIONAL MICROPHONE—A microphone that is most sensitive to sounds arriving from one direction, in front of the microphone. Examples are cardioid, supercardioid, and hypercardioid.

VALVE—The British term for vacuum tube.

VIRTUAL CONTROLS—Audio equipment controls simulated on a computer monitor screen. You adjust them with a mouse.

VIRTUAL LOUDSPEAKER—A transaural image synthesized to simulate a loudspeaker placed at a desired location.

VIRTUAL SURROUND SYSTEM—An audio reproduction system using two speakers to create the illusion that the listener is surrounded by virtual loudspeakers in a 5.1 surround array.

VIRTUAL TRACK—A sequencer recording of a single musical line, recorded as data in computer memory. A virtual track is the computer's equivalent of a tape track on a multitrack tape recorder.

VU METER—A voltmeter with a specified transient response, calibrated in VU or volume units, used to show the relative volume of various audio signals and set the recording level.

WAVEFORM—A graph of a signal's sound pressure or voltage versus time. The waveform of a pure tone is a sine wave.

WINDSCREEN—*See* Pop filter.

XLR-TYPE CONNECTOR—An ITT Cannon part number that has become the popular definition for a three-pin professional audio connector. *See also* Three-pin connector.

XY—*See* Coincident-pair method.

Y-ADAPTER—A cable that divides into two cables in parallel to feed one signal to two destinations.

Z—Mathematical variable for impedance.

INDEX

E

F

G

H

I

J

K

L

M

N

O

P

R

S

T

U

V

W

X